GAME CHANGERS

GAME CHANGERS

Stories of the REVOLUTIONARY MINDS behind GAME THEORY

RUDOLF TASCHNER

ENGLISH TRANSLATION BY BRIAN TAYLOR

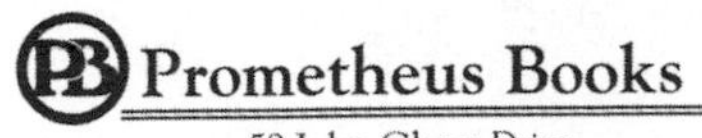

59 John Glenn Drive
Amherst, New York 14228

Published 2017 by Prometheus Books

English translation by Brian Taylor, 2017
Originally published by Carl Hanser Verlag of Munich, Germany, in August of 2015
Rudolf Taschner, *Die Mathematik des Daseins,* © Carl Hanser Verlag München 2015.

Cover image © Media Bakery
Cover design by Liz Mills
Cover design © Prometheus Books

Inquiries should be addressed to
Prometheus Books
59 John Glenn Drive
Amherst, New York 14228
VOICE: 716–691–0133 • FAX: 716–691–0137
WWW.PROMETHEUSBOOKS.COM

21 20 19 18 17 5 4 3 2 1

Library of Congress Cataloging-in-Publication Data

Names: Taschner, Rudolf J. (Rudolf Josef) author.
Title: Game changers : stories of the revolutionary minds behind game theory / by Rudolf Taschner.
Other titles: Mathematik des Daseins. English
Description: First American trade paperback edition. | Amherst, New York : Prometheus Books, 2017. | Includes index. | "Originally published as Die Mathematik des Daseins by Carl Hanser Verlag GmbH & Co. KG, Munich, 2015."
Identifiers: LCCN 2017017649| ISBN 9781633883734 (paperback) | ISBN 9781633883741 (ebook)
Subjects: LCSH: Game theory. | Mathematics—History. | Economics, Mathematical. | Game theory—History. | Existence theorems. | BISAC: MATHEMATICS / Game Theory. | BUSINESS & ECONOMICS / Insurance / Risk Assessment & Management. | PHILOSOPHY / Logic.
Classification: LCC QA269 .T3813 2017 | DDC 519.3—dc23
LC record available at https://lccn.loc.gov/2017017649

Printed in the United States of America

CONTENTS

The Players 9

Acknowledgments 11

Vienna, 2017

1: Playing with Water and Diamonds 13

Vienna, between 1870 and 1928

2: Playing with Chalk 25

Vienna, 1921

3: Playing with Numbers 37

Lyon, 1612

4: Playing with Chance 47

Port Royal des Champs, near Paris, 1655

5: Playing with Time 57

Philadelphia, 1746

Amsterdam, 1636–1637

6: Playing with a System 69

Paris and Port Royal des Champs, 1659

St. Petersburg, 1738

7: Playing with Scholars 83

Vienna, between 1921 and 1934

8: Playing with Two Cards 95

Princeton, New Jersey, 1938

9: Playing with Life and Death 109

Budapest, 1908

Princeton, New Jersey, between 1929 and 1957

10: Playing with Chickens and Lions 119

Princeton, New Jersey 1949

11: Playing with Prisoners 131

Stanford, near Palo Alto, California, 1949

12: Playing with Profit 141

Berkeley, near San Francisco, California, 1980

13: Playing with the Police 153

Vienna, 2002

14: Playing with Information 165

New York City, 1990

15: Playing with Language 175

Cambridge, between 1928 and 1946

16: Playing with Emotions 185

Ios, around 850 BCE

Barcelona, 2014

Rome, 1900

Vienna, 1786

17: Playing with Existence 195

Paris, 1662

Number Games 203

Exercises 203

Answers 210

Glossary 217

Notes 225

Index 229

THE PLAYERS

(in order of appearance)

Carl Menger von Wolfensgrün (1840–1921): Economist, founder of the theory of marginal utility and the Austrian School of Economics

Hans Hahn (1879–1934): Mathematician and cofounder of the Vienna Circle

Claude Gaspard Bachet de Méziriac (1581–1638): Mathematician, linguist, polymath, member of the Jesuit order for one year

Pierre de Fermat (1607–1665): Lawyer and amateur mathematician, invented probability theory together with Blaise Pascal

Benjamin Franklin (1706–1790): Printer, publisher, writer, natural scientist, inventor, and statesman

Nikolaus Bernoulli (1687–1759): Nephew of Jakob Bernoulli, editor of his uncle's *Ars Conjectandi*, inventor of the St. Petersburg paradox

Karl Menger (1902–1985): Mathematician, member or associate of the Vienna Circle

Oskar Morgenstern (1902–1977): Economist, director of the Austrian Institute for Business Cycle Research, invented game theory together with John von Neumann

John von Neumann (1903–1957): Mathematician, mathematical physicist, computing pioneer, invented game theory together with Oskar Morgenstern

John Forbes Nash Jr. (1928–2015): Mathematician, game theorist, inventor of the Nash equilibrium

Albert William Tucker (1905–1995): Mathematician, game theorist, doctoral supervisor to John Nash

Anatol Rapoport (1911–2007): Concert pianist, mathematician, game theorist, biologist and expert on systems theory
Karl Sigmund (born 1945): Mathematician, game theorist, mathematical historian specializing in the time of the Vienna Circle
Marilyn vos Savant (born 1946): Columnist and writer
Ludwig Wittgenstein (1889–1951): Philosopher, primary school teacher, gardener, architect, inventor of the language game
Wolfgang Amadeus Mozart (1756–1791): Composer
Blaise Pascal (1623–1662): Mathematician, physicist, man of letters, philosopher, friend and advisor to the avid gambler and soldier of fortune Antoine Gombaud, who was known as the Chevalier de Méré
Along with a host of other characters, both historical and invented, as their co-players

ACKNOWLEDGMENTS

VIENNA, 2017

The idea for this book was conceived in one of Vienna's beautiful coffeehouses, and my wife, Bianca, managed to use her irresistible enthusiasm and conviction to persuade first Christian Koth from the Hanser publishing house and then me that game theory was indeed a subject worth writing about, particularly when it is not forced into a mathematical corset but is instead addressed in a broader context. I would like to thank Mr. Koth and the considerate, generous people at Hanser for the trust they placed in me. An author could not wish for a better publishing team.

I am delighted that, thanks to the initiative of my friend and colleague, the outstanding mathematics teacher Alfred Posamentier, the renowned publishing house Prometheus decided to make the book available to an American audience. I am extremely grateful to Alfred Posamentier, as well as to Steven Mitchell and his excellent team at Prometheus, for making this possible, and I must also thank the book's translator Brian Taylor, whose experience, care, and accuracy are matched by his feel for both German and English.

I owe particular gratitude to two mathematical colleagues and friends for helping me get to know more about game theory. The first is Alexander Mehlmann, who works at math.space, the project organized by my wife in Vienna's Museumsquartier and sponsored by the Austrian government, and which allows the wider public to be exposed to mathematics as a cultural achievement. In his profound yet witty lectures, Mehlmann looks with great circumspection at mathematical game theory, the specialist field he has made his own, inspiring enthusiasm for the subject among laymen and professional mathematicians alike, including myself. The second person I must thank here is Karl Sigmund, the doyen of the Faculty of Mathematics at the University of Vienna, one of the great mathematical luminaries of our time, not merely an unerring expert on probability and game theory, but also an outstanding authority on mathematical history of the last century, whose exhibitions held in numerous venues and whose books about Kurt Gödel and the Vienna Circle have set new standards.

Above all, I am grateful to Karl Sigmund for having cast a critical eye over the majority of my manuscript, thus helping me to eliminate many—all, I hope—embarrassing mistakes I had made while writing the book. In addition to this, I was also able to enlist the help of one of Austria's top science journalists, Thomas Kramar from the daily newspaper *Die Presse,* to read through the entire manuscript. The advice given by Sigmund and Kramar was of particular value, since I decided in this book to have my heroes appear in scenes and use direct speech. Since it is a Sisyphean task to present history as it actually was, I thought it might be more amusing and perhaps even more instructive for readers if the historical facts were transformed into fictional, dramatized accounts. The use of the present tense in these lively scenes enables the reader to recognize the shift from the more matter-of-fact reports about events. Meir Shalev once said that the stories that we tell are more precise than reality, and I completely agree. In this respect, everything that is described in the book is believable and true for the very reason that I made it all up . . .

I hope that people will gain as much enjoyment from reading the book as I did from writing it. I owe the blessed position that I was in above all to my wife and to my children, Laura and Alexander, all of whom offered constant help and encouragement during every stage of the book's development.

1
PLAYING WITH WATER AND DIAMONDS

VIENNA, BETWEEN 1870 AND 1928

"This book is a masterpiece!"

Karl Menger positively shines with joy when he hears these words. He was expecting praise, but to have it lavished upon him like this by his teacher Hans Hahn is still a pleasant surprise.

No pupil, with the exception of Kurt Gödel, a curious, maverick character at the Institute, was rated as highly by Hahn as young Menger. People always had to say "young Menger" when referring to him, since his father, "old Menger," had been a professor at the University of Vienna some years before, enjoying fame as the founder of the Austrian School of Economics. If that wasn't enough, father and son both had what sounded like the same name, the only difference being the first letter—the archaic C for the father, and the modern K for the son.

In actual fact, old Menger had hoped that his son would follow in his footsteps. He himself came from the furthest reaches of the Habsburg Empire—Neu-Sandez in Galicia, a town known for its Chassidic community and Rabbi Chaim Halberstam, who taught there until his death in 1876. The son of a well-to-do family of civil servants—his father, Anton, was a lawyer and his mother, Caroline, was the daughter of a prosperous Bohemian merchant—Carl Menger came into the world in 1840. It was a small and placid world. Since

Empress Maria Theresia had unwillingly received Galicia in 1772 in exchange for wealthy and much-coveted Silesia—"She wept, but she accepted," as her victorious archrival, the Prussian King Frederick, mockingly put it—the Habsburgs had been posting doctors, lawyers, teachers, and civil servants to this dreamy land. This was what they did with all of their provinces, a custom that, while not exactly giving the regions freedom, let alone independence, did bestow on them a certain prosperity, security, and progress.

For young Carl Menger, this world was much too circumscribed. He studied law in Prague and the imperial capital Vienna, where he then settled down as a journalist. He wrote feature articles, first for the *Lemberger Zeitung*—Lemberg (now known as Lviv) was the Galician capital—and later for the *Wiener Zeitung*. He was fond of writing novels and comedies for serialization, occasional works that he produced while following his true interest, the study of law and political economics. He enjoyed the acquaintance of Count Richard von Belcredi, the minister of state at the time, who familiarized him with economic issues. This enabled Carl Menger to enrich the *Wiener Zeitung* with his market analysis—the start of his deep interest in the laws of economics.

There was one paradox of economic theory that occupied the young lawyer and economic journalist in the months after gaining his doctorate: the paradox of value, also known as the diamond-water paradox. Nobody can survive without water. This makes it a highly valuable commodity. Very few people truly need diamonds. This makes them intrinsically almost worthless. And yet people pay horrendous sums of money for diamonds and almost nothing for water.

Adam Smith, the eighteenth-century founder of political economics, thought he could solve this paradox by differentiating between "value in use" and "value in exchange." The value in use of water is very high, since everybody needs it. Its value in exchange, on the other hand, is very low. In contrast, the value in exchange of diamonds is very high, hence the high prices paid, although their value in use is low. This rudimentary explanation is not particularly satisfactory, however, since it sheds no light on *why* there is a difference in the value in exchange.

Even before Adam Smith, the Scottish banker John Law had stated, "Water is of great use, yet of little value; because the quantity of water is much greater than the demand for it. Diamonds are of little use, yet of great value, because the demand for diamonds is much greater than the quantity of them." With this statement, he may well have found the pivotal point that might help to solve the paradox.

By way of explaining the diamond-water paradox, Carl Menger comes up with the idea of a farmer who owns five sacks of wheat. The farmer considers the first sack of wheat as indispensable for life, since he uses it to bake his bread, and so he will not die of hunger. With the wheat in the second sack, which is still valuable for him, he bakes even more bread. This gives him and his family strength. With the wheat in the third sack, which is not particularly valuable, he can feed the animals in his stable. He puts aside the wheat in the fourth sack for sowing the following year. The farmer doesn't actually need the wheat in the fifth sack; he uses it to distil grain schnapps.

If one of the farmer's five sacks were to be stolen, what would he do? If his use of the wheat were always the same, he would divide the wheat in the remaining four sacks into five equal heaps and use each heap for the same purpose as with the original sacks: the first heap for the bread necessary for survival, the second heap for the bread to give him strength, the third heap to feed the animals, the fourth heap for seed, and the fifth for making schnapps. But no farmer would do such a thing—farmers are clever souls, as we know. Rather, he would use the remaining four sacks as described before, omitting only the unnecessary distilling of schnapps.

Indeed, farmers are much too clever to let people steal from them. But the farmer could sell the fifth sack, instead of using it to make schnapps that he and his family might never drink. And what would the farmer consider to be the right price? Carl Menger has the answer: it is the price that the farmer would pay to buy the fifth sack in order to use the wheat to distil the schnapps.

It is not the primary or the secondary use of wheat that determines the price, but rather its marginal use—the use that the farmer makes of an extra sack on top of those he already has in storage. That, says Carl Menger, is why water is so cheap: another liter of water on top of the overflowing abundance already available is viewed as insignificant. Only in the middle of the Sahara can water's worth be measured in diamonds.

The economic theory developed from these simple premises brought Carl Menger fame in the world of economics and politics, and he became one of the most influential figures in the Habsburg monarchy. The University of Vienna named him first an associate professor and later a full professor at the Faculty of Law and Political Science, and he even came to the notice of Emperor Franz Joseph himself, being accorded the honor of spending three months expounding the features of his economic theories to the monarch. He was appointed as the private tutor for the emperor's eighteen-year-old son Rudolf, and the two of them spent two years traveling the length and breadth of Europe. During this time, and in the years that followed until Rudolf's death, it seems that Menger became friends with the highly gifted and sensitive young man, awakening in him an interest in modern, liberal governance. But all the hopes that the Liberals pinned on the young crown prince were dashed with the shots the thirty-year-old Rudolf fired in 1889 to kill his young lover, the seventeen-year-old Baroness Mary Vetsera, and himself.

It was in all probability thanks to his close links to the monarchy that Carl Menger's son Karl, born in Vienna in 1902, was recognized as a legitimate child. For Karl's mother, Hermine Andermann, was Jewish. The Catholic father and the Jewish mother were unable to get married in a land where only marriage ceremonies carried out by the Church or the Synagogue were recognized. The two lived in a union akin to matrimony, in what was then known as a *marriage sui juris.* Children resulting from such unions were considered to be illegitimate, which signified social ostracism, and Carl Menger took early retirement shortly after the birth of his son in order to protect his family from gossip. He was, therefore, all the more grateful to the monarchy for fulfilling his request to annul his son's status as an illegitimate child.

Menger's departure represented a severe loss for the university, one only exacerbated by the fact that, even though no longer a professor there, he remained in contact with his pupils. Felix Somary, whom Menger appointed as his assistant at the tender age of eighteen during his last years of active service, writes in his memoirs:

> At that time, the University of Vienna stood at the forefront of the world's schools of national and political economics. Carl Menger, the leading theoretician, followed by his eminent pupils Böhm-Bawerk and Wieser; Philippovich, the brilliant investigator of productivity; and Inama-Sternegg, the first economic historian, were all there—a unique collaboration of remarkable personalities. The discussions held in the seminars were of an exalted standard, since there were also exceptional talents among the students of my time, such as Schumpeter, Pribram, Mises, Otto Bauer, Lederer and Hilferding. Not one of them would end up staying in Austria.[1]

Despite all his success, one imagines Carl Menger as a melancholic man. His academic achievements were fiercely contested by the German "Historical School of Economics," specifically by its main representative, Gustav Schmoller. The accolades he received, his title of *Hofrat* (court counsellor) and admittance to the *Österreichisches Herrenhaus* (Austrian House of Lords) all meant little to him, a man who never made use of his aristocratic title "von Wolfensgrün." In the renunciation of liberalism, he saw the descent into disaster, and he felt vindicated in this view after the outbreak of the First World War and its catastrophic effect on the Habsburg Empire. "The World of Yesterday," as Stefan Zweig had called it, lay irrevocably in ruins, and the dream of a better, prosperous world with a firm economic foundation had vanished forever.

Given such a father, who had, after all, reached the age of sixty-two at the birth of his son and only child, young Karl's childhood was certainly not unencumbered by misery, no matter how financially secure it was. With the commanding image of his father before his eyes, the boy set himself the highest standards from his early years. He attended one of the city's best grammar schools, in the wealthy district of Döbling, between the Vienna Woods and the city center, and felt constantly obliged to get only top marks, ideally the very best in the school. But this was no easy undertaking, since Richard Kuhn and Wolfgang Pauli were fellow pupils of his, just two years above him. The highly gifted Richard Kuhn went on to study chemistry, developed chromatography to the extent that it was suitable for chemical analysis, and was awarded the Nobel Prize for his work on carotenoids and vitamins. Wolfgang Pauli's brilliant mind was even more impressive. He later became known as the "conscience of physics" because his critical eye was always able to separate the wheat from the chaff when it came to all the hypotheses developed by Bohr, Heisenberg, Dirac, and the other notable names of quantum physics. He received the Nobel Prize for his discovery of the Pauli exclusion principle, which helped physicists to find an explanation for the existence of the periodic table of chemical elements.

It is certain that Karl Menger attempted to measure his intellect against Pauli's. Pauli delighted in his intellectual powers. Once, when the physics teacher made a mistake on the blackboard and couldn't find it, despite spending minutes looking for it, Pauli simply grinned devilishly from ear to ear until the teacher finally called out to his prodigy, much to the amusement of the class: "Now Pauli, tell me where the mistake is. I know you've long since found it." After doing his *Matura*, the Austrian school-leaving exams, Pauli published a scientific article on the general theory of relativity that impressed even Einstein himself. And after just a few terms studying physics in Munich, he wrote a long piece on the theory of relativity for the *Encyclopedia of Mathematical Sciences*, an article that was published as a work in its own right in 1921, received lavish praise from Einstein, and came to be viewed as a classic.

Karl Menger, Pauli's schoolmate and admirer, was convinced that physics was a subject against which one could measure one's intellectual capabilities, but it was no easy task to persuade his father, who had been living like a recluse since the end of the war in 1918, to share the same opinion. We can almost hear Carl Menger ranting: "First you wanted to go into the theater, which thank God I was able to talk you out of, and now you have turned your mind to this pauper's subject?" Karl had indeed wanted to measure himself not only against Pauli but also against his fellow student Heinrich Schnitzler. Schnitzler's father, Arthur, was one of the leading German-language dramatists of the time, and Heinrich himself had resolved to become an actor and film director, since which his admiring friend Karl had been thinking about writing plays. Drafts of a drama featuring the fabled medieval Pope Joan as its main character lay in the drawer of his desk.

Karl tries to explain: "Forget my theatrical ambitions, Papa, and believe me when I say that studying physics is the right choice. I believe I have a gift for the subject. I am seriously interested in the principles of natural science and, in Goethe's words, in 'what binds the world together at its core.' Besides, it wouldn't be wise to try my hand at economics—I could never compete with you and nor would I want to." His father mumbles something incomprehensible and points his son out of his office. Karl Menger is now free to study physics.

On Währingerstrasse, which runs from the center of Vienna to the outer districts of Währing and Döbling, the large gray building at numbers 38 to 42 housed the institutes of chemistry, physics, and mathematics. Here were the state-of-the-art laboratories, the seminar rooms and lecture halls. The place's best days were behind it, however, since the new republic was devoid of modern industry after the war. Following the collapse of the monarchy, the major industrial centers, above all those in Bohemia and Moravia, had been lost to Czechoslovakia and other successor states. With a capital city of two million people, among them far too many civil servants, and a hinterland still very much dependent on an outdated agricultural system, an impoverished country full of war invalids, widows, and orphans and plagued by the Spanish flu pandemic that was rife at the time could not possibly make any great advances. Major support for scientific research was unthinkable. People were glad enough that the university was able to continue operating after a fashion and could retain its staff despite the meagre salaries. Perhaps Pauli is right to matriculate at Ludwig Maximilian University in Munich, young Menger must have thought to himself as he enters the large dusty mathematics lecture hall on the lower ground floor. But it is still possible to change universities after a couple of semesters. At least for now, he doesn't have the heart to leave his parents, especially his aged father, alone in the desolate surroundings of post-war Vienna.

The general theory of relativity, with which Menger's schoolmate Pauli had already made a name for himself, was expounded up on the fourth floor of the same building by the brilliant professor of physics Hans Thirring, a friend of Einstein and a fellow pacifist. The topic was not included in the beginner's lectures on physics, however, which new students had to attend on the lower floors. In addition, it was compulsory to attend mathematics lectures on the ground floor, these being designed to provide the tools for dealing with theoretical physics. This order was also reflected in the timetable: the mathematics lectures took place at eight and nine o'clock, with the physics lectures following at ten and eleven. And so we can observe young Menger descending the short stairway to the mathematics lecture hall at eight o'clock and joining the handful of people already there.

A triumvirate of professors, all three of international standing, occupied the research and teaching posts at the mathematical institute. As a young man in Göttingen, the bastion of mathematics at the time, Wilhelm Wirtinger, originally from Ybbs, a small town one hundred kilometers west of Vienna, had studied at the feet of the famous scholar Felix Klein and established himself as a gifted and versatile mathematician. His tall, gaunt figure, the neatly trimmed moustache on his oval face, his high forehead, and the dignity with which he carried out his duties all lent him a commanding appearance. And Hofrat Wirtinger—he was the last of the Viennese professors to have received the title of court counsellor—gave lectures of the highest level, extremely demanding and only for students who had carefully prepared for them.

Born in Elze in Lower Saxony, Philipp Furtwängler was another Göttingen alumnus to come to Vienna as a professor of mathematics. He was the cousin of the conductor Wilhelm Furtwängler and specialized in number theory, the field that Gauss called the "Queen of Mathematics." Gödel praised his lectures as the best he'd ever heard, and they were indeed unique. A disease meant that Furtwängler was paralyzed from the neck down; he had to be carried down the steps to the lecture hall by assistants, and he sat in a chair in front of the board during the lecture and dictated what was to be written on the board to one of his assistants. Gödel, since childhood a hypochondriac of the first order and consumed by the delusion that he was on the verge of death, had actually intended to study physics like Karl Menger, but then decided to switch to mathematics on account of Furtwängler's lectures. Perhaps also for the following reason, which he may have arrived at with his bizarre, logical conclusions: "Furtwängler, that's obvious, is truly ill. I, Gödel, just imagine I am—though you never know. However, if Furtwängler, though truly ill, can grow so old, then concerning oneself with mathematics must be conducive to a long life. If I should genuinely be ill, therefore, it makes sense to devote myself to studying mathematics, since it has a life-prolonging effect."

It was the third member of this illustrious triumvirate, Hans Hahn, who inspired Karl Menger to switch from physics to mathematics. The reason behind this change was not as eccentric as with Gödel and was quite comprehensible for normal-thinking people: Menger was fascinated by Hahn not so much because of the brilliance with which he lectured but much rather because what he had to say was completely different from what one was used to hearing from mathematicians, with their equations and never-ending formulae. In his lectures, Hahn focused on the absolutely essential: the aim was to investigate those terms that sounded particularly simple.

"What is a curve?" Hahn asks the bemused audience in the lecture hall, and says nothing more for a few seconds. With a mischievous smile, he interrupts the silence he has created. "If you believe, ladies and gentlemen"—a few women had already plucked up the courage to take on such a typically male-dominated subject as mathematics, some of them with considerable success, like Eleonore Minor, later to be Hahn's wife, or Olga Taussky—"if you believe that this question is too simple, then just forget it. Continue to wander, like so many others, along the well-trodden paths of the classical mathematics of the last century. But only attend my colleagues' seminars, not mine. You see, in my course this summer term, I am only going to look into this one question: What is a curve?"

Young Menger is electrified by what he hears. It goes without saying that he wishes to take part in the seminars, even though he is just beginning his studies.

After the first lesson, in which Hahn talks about the various attempts by mathematicians to give a precise definition of a curve, with none of these suggestions being fully satisfactory, we can see Menger walking back home, confused and delighted in equal measure, having been privy to the discussion of a problem for which a solution must be found using crystal-clear, logical reasoning. He locks himself in his room, forgets everything around him, and focuses solely on the attempt to put his thoughts as precisely as possible down on paper. Hours later—the wastepaper basket is overflowing and Menger is completely exhausted—he believes that he has sighted light at the end of the proverbial tunnel.

"Perhaps it's just an oncoming train." Menger's school friend Otto Schreier, who is a year older than him and now attends mathematics lectures with him, attempts to bring Menger, who is full of enthusiasm about his own ideas, back down to earth. The two of them meet up where people in Vienna always meet—at the coffeehouse. Young Menger tells the skeptical Schreier about the thoughts he has spent hours poring over the day before. Schreier tries to convince him to see sense: "Just consider: a noted mathematician like Hausdorff in little Greifswald has investigated the most outlandish examples of curves, and he doubts that an exact definition of the curve will ever be found. Hahn probably hasn't yet mentioned this in his seminar."

"I'm going to Hahn's office hours," declares Menger, convinced that the solution he has conceived will receive the professor's blessing, and Schreier wishes him luck.

Hahn does indeed seem rather underwhelmed at first, when the youthful Menger enters his office and reveals to him that he believes he can answer the question about the nature of a curve. But when the impassioned young man tries to formulate his ideas, Hahn recognizes the prodigious talent that Menger possesses for answering the kind of questions he loves. "Keep going," he prompts him. "Your approach is very promising. Moreover, I believe it leads somewhere." When Menger finally leaves the room, he feels as though he is walking on air. He now knows that he is destined for mathematics.

Hahn supported Menger to the best of his ability. In just a short time, Menger completed his studies with Hahn and was awarded a doctorate, and Hahn got him a job as the assistant to one of the world's most eminent mathematicians, the Amsterdam-based Luitzen Egbertus Jan Brouwer. And when Menger returned to Vienna, not yet even twenty-six years of age, he was appointed to a professorship of geometry, and the manuscript of his book was already almost finished. *Dimension Theory* was to be the title.

With a newly printed copy in his hand, he knocks at the door of Hahn's office at the Institute and proudly hands it over. "This book is a masterpiece": Hahn is not the only one to lavish such praise upon the work.

2
PLAYING WITH CHALK

VIENNA, 1921

"What is a curve?"

We return to the year 1921, to the moment when Karl Menger, only newly enrolled at the university, is listening intently to Hans Hahn speaking. In the small seminar room, the professor starts off by posing the question with the same penetrating voice as in the large lecture hall, before repeating it for theatrical effect: "What is a curve?" He draws a big treble clef on the board. "All of you will say that this is a typical curve. And you will be right. But showing an example that everybody can understand does not provide a definition. The word 'definition' contains the Latin 'finis,' which means 'boundary.' A definition sets a distinct boundary. In this seminar, we are trying to set such a distinct boundary so that everything that is contained within it can justifiably be called a curve and nothing that is outside this boundary deserves the name."

Hahn draws a straight line on the board. "Some of you will think that we shouldn't talk of a curve here, because curves always have a bend. The *Fiaker*"—the drivers of the Viennese horse-drawn carriages—"see it like this too. When they drive along the dead-straight avenue in the Prater park, they would never say that they were moving along a curve. For them, a curve means that they have to steer their horses to the right or left with the reins. But we here," Hahn gestures with his powerful right arm so as to include his listeners, "are mathematicians and not *Fiaker.*" There are smirks in the audience. "We will also permit a straight line without a bend to be a curve. That makes a lot of sense." Hahn uses his chalk to add a semi-circle to the end of the line already drawn. "All of you will call this hooked shape a curve. The straight line that I drew before is part of this curve. After all, it would be unreasonable if we were suddenly to say that this straight part of the curve is not a curve."

"But not every part of a curve has to be a curve," objects a presumptuous older student.

Hahn pricks up his ears, delighted. "Yes, carry on," he says, to encourage the student's interruption.

"I wouldn't refer to a single point of the curve as still being a curve," the student continues, "and when I look at two separate parts of a curve, it may be that both of the parts are still curves. But viewed as a whole, the two parts may no longer constitute a curve."

"Excellent," answers Hahn, beaming at the student. "That enables us to establish at least two features of a curve: first, it must always consist of more than just a single point. And if we take two points of the curve, it must be possible to link them both along the curve." After what seems like a significant pause, he continues: "But how exactly is this to be understood—that we 'link two points along the curve'?"

Menger listens, spellbound. It is only very faintly that his conscience pipes up, asking in his father's voice: "You really want to study *that?* You want to devote your life to such academic questions, so far removed from reality? Aren't they just fantasies, pipe dreams, and intellectual bubbles? Law, medicine, or even technology, for all I care—these are tangible courses of study that will help you achieve something sensible. I won't even mention economics. Have you really given this enough thought?"

As Hans Hahn draws a zigzag on the board with his chalk, Menger dispels these quietly insistent objections. He is all too intent on not missing a single detail of the lecture. "Some mathematicians don't allow any vertices in curves," he hears Hahn explain. "A rounded-off rectangle is in their view a curve, a kind of oval shape, but the corner of a proper rectangle is not. I do not wish to go along with such mathematicians who insist on only 'smooth curves.' After all, the edge of a rectangle with its four vertices arises from the overlapping of radius corners of rounded-off rectangles, when the radii of the circles at the curvatures get smaller and smaller. If I wish to accept such boundary curves of smooth curves as 'curves' in their own right, then I must call a zigzag line like the one I have just drawn a curve, too."

Perhaps Menger has not yet consciously grasped the fact, but he must surely have realized with sudden intuition that he is now in a position to respond triumphantly to his conscience and its probing protests spoken in his father's voice. For he knows such zigzag lines as those that Hahn has just drawn on the board from the business pages of the newspapers, as well as from his father's very own work and that of his colleagues and pupils—they are the graphs of companies' share prices.

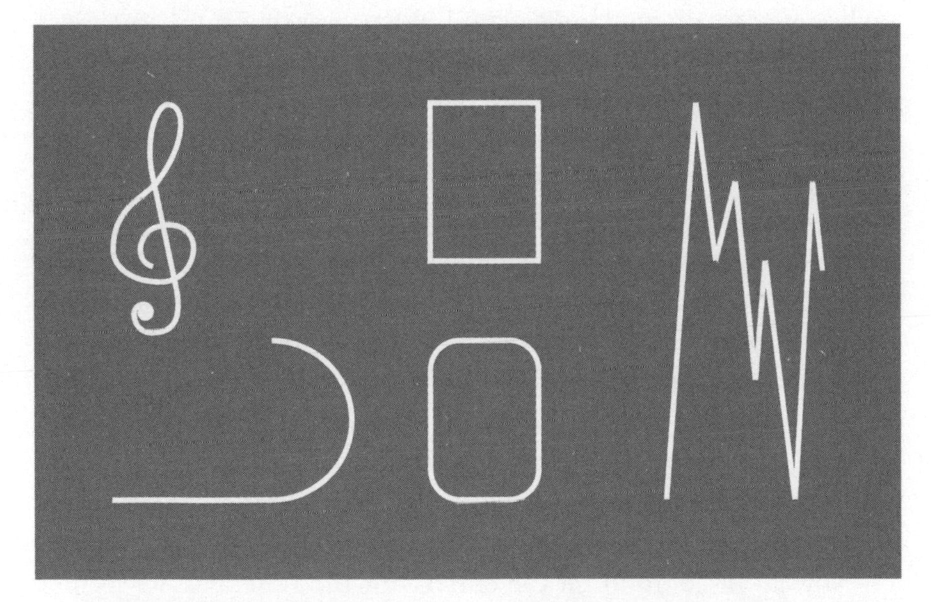

Figure 2.1: Hans Hahn's drawings showing a treble clef at the top left as a smooth curve; a straight line joined to a half-circle at the bottom left; the outline of a rectangle as a curve with four corners in the top center; the outline of a rounded-off rectangle in the bottom center; and a zigzag line on the right.

Even then, in the early twentieth century, the odd diagrams that showed the development over time of share or bond prices played an important role for brokers and investors. The time axis stretches to the right. Vertically, from each point along the axis, the price of the security in question is recorded: high up, when the price reaches stratospheric levels, or low down, when it reaches rock bottom. The structure of points that so arises forms the archetypal curve. It may not be a smooth curve like the treble clef that Hahn drew on the board at the beginning of the seminar, but it is just as much a curve as the zigzag about which he is now speaking. And it is a curve on which many people's eyes are fixed, for with curves such as these there is an awful lot of money at stake. Shareholders gaze in horror when the curve plummets, while boundless jubilation rings out when it suddenly shoots almost vertically upward.

Zigzag curves are not some figment of the imagination, dreamt up by out-of-touch mathematicians hidden away in their ivory towers. The zigzag curves of the stock exchange are the very essence of the stock market. The very existence of individuals and families, companies and countries, depends on the way they develop. It is therefore reasonable to ponder their peculiar nature.

After all, they are indeed peculiar. Let's take a look at the share price performance of a listed company, featuring the usual fluctuation, over the period from 1871 to 1921. Some of the dramatic falls and sudden increases can be linked directly to the historic events that happened at the same time as these changes: from the 1873 World Exhibition in Vienna or the annexation of Bosnia in 1908 to the disruption of the First World War. But not everything can be explained by such major events, and the constant fluctuation can also be observed in more sedate political times.

Now let us zoom in on the curve—for the time axis, let us look at the period between 1916 and 1921 and record the company's share price development for this time. There is no change in the fluctuation observed before. The steep lines from before are not flattened out and do not change more serenely. In actual fact, they are frayed to form an unstable zigzag shape: the curve may well give a different impression from the previous one when viewed from afar, but it retains its erratic fluctuation. Exactly the same thing happens when we concentrate solely on the period between January 3, 1921, when the stock exchange opened for the first time in the new year, and Friday, March 4, 1921. Where for 1921 in the previous graph we had assumed a gentle, shallow progression, we suddenly see in the enlarged version (the vertical axis is correspondingly enlarged and thus more finely scaled) significant variations. And even when we only take into account the week from February 28 to March 4, there is no change to this constantly erratic fluctuation.

How simplistic in comparison is the smooth curve of the treble clef that is still emblazoned on the blackboard of the seminar room. Anybody who wishes to represent with complete precision the share price performance of the company in question is destined to failure. The first zigzag line in particular, drawn for the period between 1871 and 1921, is almost criminally imprecise. But none of the enlarged sections provides an exact illustration of the share price performance, either. Not one of the infinite number of spikes that change the progression literally from moment to moment can be properly captured by a sketch, no matter how accurate one attempts to be.

If one wanted to represent the graph of the share price performance with perfect precision, one would be sure to fail, rather like Laurence Sterne in his biography of Tristram Shandy: since the author wants to mention every single detail of his hero's life, no matter how small and insignificant, he never really gets much beyond Tristram Shandy's birth, despite the nine volumes that make up his work, each with forty chapters. A similar fate would befall a pedantic illustrator aiming to record the exact curve representing the share price performance between 1871 and 1921: he would remain stuck on Monday, January 2, 1871.

The length of the curve of the treble clef is relatively easy to determine. The best thing is to draw points close to each other along the curve and replace the sweep of the treble clef between two neighboring points by a straight line connecting them. One then merely has to add up the lengths of these short individual straight bits. This may not give the exact length of the curve itself, instead measuring a slightly shorter stretch, but when one plants the interspersed points very close to one another, the measurement of the treble clef's length is accurate enough.

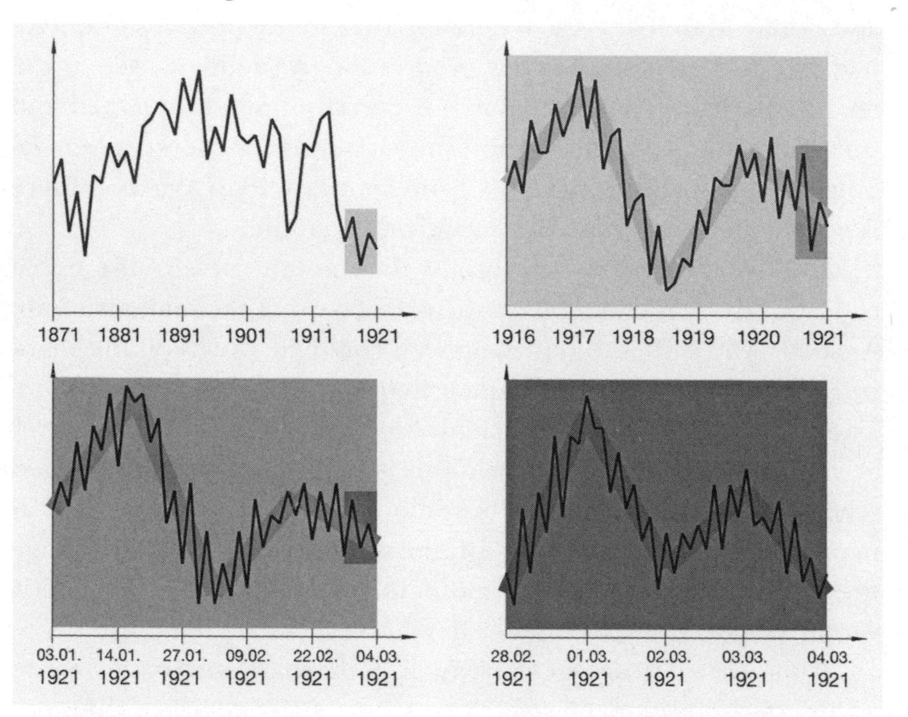

Figure 2.2: At the top left, a graph showing a company's share performance over a long period. The section highlighted in the light-gray rectangle is shown in close-up at the top right. The section highlighted in the medium-gray rectangle is shown in close-up at the bottom left. The section highlighted in the dark-gray rectangle is shown in close-up at the bottom right. Each time the graph is enlarged, what seems to be a section composed only of four lines frays further to form a zigzag line.

When it comes to the erratic zigzag curve of the share price performance, this method of measuring length is pointless. The curve has no "true length." For it would be a matter of linking an infinite number of points, with no reduction in this abundance of points even in the smallest section of the curve. It is simply futile to talk of the "length" of the curve that represents the share price performance. Even if one says that it is infinitely long, that doesn't signify much. What does "infinitely long" mean?

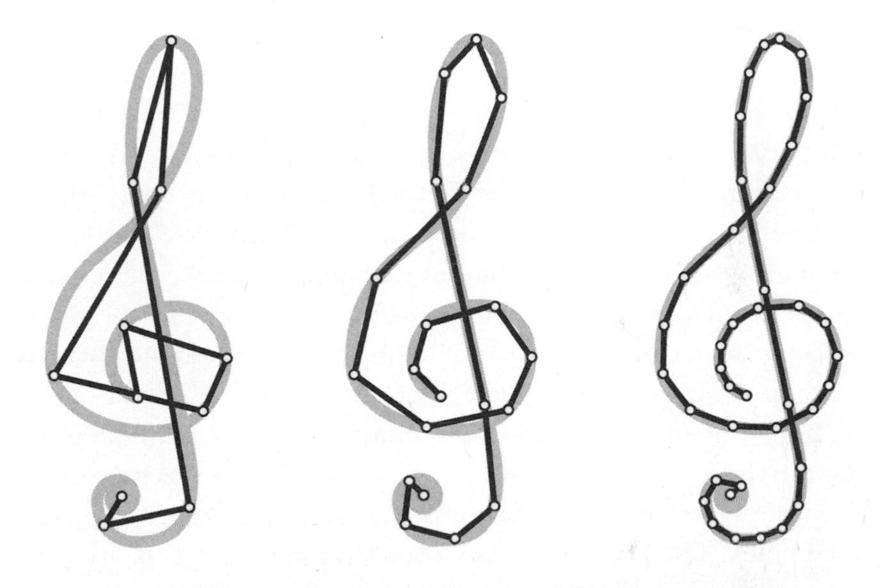

Figure 2.3: The length of a smooth curve—for example, the treble clef drawn in gray—can be measured approximately by connecting points along the curve with straight lines and adding up the lengths of these lines. On the left, very few points have been drawn; in the middle, there are more; and on the right, there are many points connected by straight lines. The total length of the straight lines increases from left to right, providing an increasingly accurate measurement of the smooth curve's length.

These are Otto Schreier's thoughts as he sits in the coffeehouse telling his school friend Menger about Felix Hausdorff. Hausdorff's deliberations led to the idea that the curve representing the share price performance is perhaps not even one-dimensional. At least, not one-dimensional in the way that one imagines a line, whether straight or curved, to be. Such a curve is certainly not one-dimensional in the sense that one can pass along it like the *Fiaker* driving along his route. On the other hand, it is also not two-dimensional in the sense that it fills a surface area. Perhaps the "dimension" of the share price performance curve is a "fractional number," lying somewhere between one and two?

Menger has a different approach to the problem from Hausdorff: his own method is systematic. It is clear that a single point, or individual spaced-out points, is zero-dimensional. A line is one-dimensional, and the curve of the treble clef is also one-dimensional, because both curves begin and end with a zero-dimensional point. A square is two-dimensional, because it is bordered by four one-dimensional lines. Similarly, the two-dimensional circle is bordered by the one-dimensional circumference. The three-dimensional cube is bordered by six two-dimensional squares; likewise, the three-dimensional sphere is encompassed by its two-dimensional surface. In this way, step by step, one can master the integral dimensions, if not the "fractional" ones. The border of any given shape must always have one fewer dimension than the shape itself. The question does arise, however, as to whether this idea can be applied to much more complicated shapes than individual points, lines, smooth curves, squares, circles, cubes, or spheres.

This is the approach that, during Menger's talk with Hahn a week later, the professor refers to as "promising" and even "leading somewhere." And this is why, seven years later, Menger calls his book, in which these thoughts are developed into a theory, *Dimension Theory.*

There is another consideration when it comes to share price performance that neither Hausdorff nor Menger thought of, because they only had in mind the geometrical image: How do these incessant fluctuations actually arise?

All of a sudden, it is rumored that a certain Prince Dolgorukov in far-off Yekaterinburg has purchased a very sizeable number of shares in a company. In a flash, the share price shoots up. Then an unnamed but reliable source spreads the news that the report is false. The share price falls. The rumormongers waste no time in suggesting to the stock exchange operators that this source was paid by

Dolgorukov to spread uncertainty. Some people believe them and buy shares; others remain skeptical and sell their holdings, each of them sooner or later. None of this can be rationally explained—in other words, the fluctuation materializes as though by chance.

When a competent *Fiaker* is steering his carriage, he knows where he is going and guides his horses along the path of a smooth curve. The erratic zigzag curve of the share price performance, on the other hand, suggests the route of a complete drunk as he staggers along the street after the bar has closed. One never knows in advance whether he will suddenly reel to the right or left. If he is so drunk that, like the share price performance graph, he changes direction practically every second, then he won't get far from the bar, even though he should be aiming for home.

Menger had no idea at the time that, twenty years before, a young Frenchman called Louis Bachelier had described the progression of the share price of listed companies as a random walk being guided by chance. Bachelier based this idea, however, not on the lurching motion of a drunkard but on microscopic random movements that the Scottish botanist Robert Brown had described decades before. Brown studied a drop of water under a microscope and observed how tiny pollen grains moved with jerky irregularity on the surface. He himself believed that he had discovered a form of life. In reality, however, this movement, now known as Brownian motion, is caused by the water molecules that, due to their thermal motion, strike the pollen grains from all sides in no particular order, thus causing the pollen to jig from side to side. The reeling of the drunkard on the street, the jerking of the pollen grains on the drop of water, or the unpredictable fluctuation of share prices—from a mathematical point of view, all of these are the same.

Bachelier's doctoral thesis, submitted in Paris to Henri Poincaré, a giant of the mathematics world, was based on this idea. Poincaré had little interest in students. He deigned to examine not even ten of them; Louis Bachelier was the first of the chosen few. He had to defend his thesis before three eminent professors at the Sorbonne:

Paul Appell and Joseph Boussinesq, both already getting on in years, and his doctoral supervisor, Henri Poincaré, not yet forty years of age but with a knowledge of his subject that put his older colleagues very much in the shade. Poincaré was highly impressed by Bachelier and his ideas, but it seemed that it was still too much effort for him to make a serious attempt to get Bachelier a suitable post at the university. Left to his own devices, Bachelier had to make do with a few badly paid teaching jobs. Even when he finally had a chance of obtaining a fixed professorship in Dijon in 1926, his hopes were dashed by a scathing report by his contemporary Paul Levy, who was already safely ensconced in a professorship in Paris. Only much too late did Levy apologize for his prejudice and the false evaluation in his report; it made no difference, and only after Bachelier's death did the theory arising from his doctoral thesis gradually become known and acknowledged for its groundbreaking nature.

In a sense, one can say that Menger, when he heard Hahn's talk and gathered his first thoughts about it, was examining the zigzag curves of the share prices phenomenologically, that is to say based on their appearance. Bachelier, on the other hand, studied them substantially (i.e., based on their nature), because he posed the question as to how they arose.

It would be unfair to close this chapter without at least mentioning that, besides Hahn and his pupil Menger on the one hand and the lone wolf Bachelier in faraway Paris on the other, there were others who were plowing the same field in their research. We can take a look at some of them:

Although he didn't know Bachelier, Albert Einstein—then, in 1905, a level three technical assistant at the Bern patent office—was also working on the phenomenon of Brownian motion, which he was able to analyze in the manner of the great Viennese theorist Ludwig Boltzmann. His paper on Brownian motion was one of the four masterpieces that he produced in that one year, the so-called Annus Mirabilis of Albert Einstein. The three other articles concerned the interpretation of light as consisting of particles, called *photons*, with these possessing highly unusual properties (Einstein received the Nobel Prize for this paper), the special relativity theory, and the discovery of the famous formula $E = mc^2$.

Without knowing either Bachelier or Einstein, Marian Smoluchowski came to the same conclusions as Einstein at the same time as his soon-to-be-celebrated counterpart. Smoluchowski came from a prosperous family with Polish roots who lived in Vorderbrühl, a pretty little place some distance from Vienna. He studied physics in Vienna and, thanks to his outstanding results, received his degree "sub auspiciis Imperatoris," that is to say, at a special ceremony attended by the emperor or a representative appointed by him. After a few years of teaching and traveling in Paris, Glasgow, and Berlin, he finally obtained a physics chair in Lemberg (Lviv), where he joined the Polish Copernicus Society of Naturalists: the Habsburg Empire counted more than a dozen different nationalities and the Poles were well represented in Galician Lemberg. Smoluchowski wrote his groundbreaking paper on the nature of Brownian motion in 1906, when he was president of the Society. He moved to Krakow shortly before the outbreak of World War I and died in the dysentery epidemic there a year before the war ended.

And finally, without knowing Menger and separately from the findings of Bachelier, Einstein, and Smoluchowski, the young Russian mathematician Pavel Samuilovich Urysohn devised practically the same "dimension theory" as young Menger. Just as Menger was lured away from studying physics to mathematics by Hans Hahn, so the same thing happened with Urysohn: Dmitri Fyodorovich Egorov and his colleague Nikolai Nikolaevich Luzin, who both taught at Lomonosov Moscow State University, inspired Urysohn to provide as comprehensive a definition as possible of what is understood as a topological dimension.

In truth, even before Menger and Urysohn, the Dutchman Luitzen Egbertus Jan Brouwer in Amsterdam had already carried out the important preliminary work for the dimension theory that the two talented young men founded independently of each other.

Menger realized this when he spent two years acting as Brouwer's assistant, and Urysohn also realized it when he called on Brouwer during a study tour that took him via Göttingen—where the famous David Hilbert was teaching—and Bonn—where the brilliant Felix Hausdorff was working after leaving Greifswald—to Amsterdam. This study tour came to a tragic end for Urysohn: he and his fellow student Pavel Sergeyevich Alexandrov holidayed in Brittany and decided to go swimming in the Atlantic in the late afternoon of August 17, 1924. The strong current dragged the two friends out to sea and, while Alexandrov was just able to make it back to shore, only Urysohn's dead body was recovered.

Menger, who first went to Amsterdam in 1925, did not therefore get to know Urysohn personally. Brouwer and Menger, the professor and his assistant, held each other in the highest regard to begin with. But a few months later, after an intense period of work together, they fell out. On the one hand, there was the older scholar, a lean, grave man with a wrinkled face, more like a priest than an academic teacher, inapproachable despite his perfect courtesy and unshakably convinced of his sometimes eccentric views. On the other hand, the young famulus, ambitious and eager for knowledge, blessed with quick perception, harsh in his judgment—something he had learned from Pauli—and ultrasensitive to the faintest criticism. When there was a disagreement as to which of the two had actually been the first to recognize the significance of an idea, the civil relations that they carefully maintained could not prevent their growing discord.

Menger was glad when, with the first copy of his *Dimension Theory* in his hands, he was able to escape Amsterdam and the cold atmosphere he now felt there and occupy the newly vacated chair of geometry at the University of Vienna. Hans Hahn had successfully campaigned for Menger's return at the Ministry of Education. Once again, we can see the scene where the young man, just returned to his home city, presses his book into his former teacher's hand and Hahn, after reading it, gives him the greatest possible compliment: "This book is a masterpiece."

3

PLAYING WITH NUMBERS

LYON, 1612

"Let me play once more!"

"Monsieur, I warn you: you have already lost four times. You will also lose the next round, I can assure you of that. I don't want to take all of the money that you have on you. I beseech you: let us finish the game!"

"Out of the question. I want to play one more round. Luck must smile upon me at least once."

Claude Gaspard Bachet de Méziriac finally gives in and consents to play against his naive aristocratic opponent for a fifth time. The game he has invented and named "Think-of-a-number" is a very simple one: no counters are needed, no board, no dice—just two compulsive gamblers. The first person says a number from one to ten. The second person then takes a number from one to ten, adds it to the first number and says what the total is. The first person again adds a number from one to ten and says the total. The two players take it in turns like this until one of them can call out the number one hundred. Whoever says one hundred has won the game.

"Good, allow me to begin," says de Méziriac. "I say one."

"Five," shoots back his opponent.

"If you say five," mutters de Méziriac, "I'll add seven and say twelve."

"Thirteen," says his opponent. "I'll only add one."

"If you add one, I'd like to add ten this time," answers de Méziriac. "I'll say twenty-three."

"Now I say twenty-seven."

"I'll counter with thirty-four."

"Forty-four," says de Méziriac's opponent, sensing victory.

"Let me think," murmurs de Méziriac, and then calls out after a few seconds, "Forty-five."

"Forty-eight."

"Fifty-six—I'll add eight to forty-eight," declares de Méziriac.

"Now it's getting close," muses his opponent and calls out: "My next number is sixty."

"If you say sixty, I'll say sixty-seven."

"Seventy," calls out de Méziriac's opponent even louder, and continues exultantly: "You'll see—next, I'll say eighty, then ninety, then one hundred."

Silence reigns for a few heartbeats, until de Méziriac whispers, "Seventy-eight."

"Eighty," says his opponent, trembling with triumphant anticipation.

"Eighty-nine," answers de Méziriac coolly, and goes on: "You see—I have won. Whatever you say now, it must be at least ninety and at most ninety-nine. And no matter what number you say, I can then reply with one hundred."

"You're a devil!" the man snaps at de Méziriac, throws down the 1,600 livres—the first round began at 100 livres and the stake was doubled with each round—and leaves the salon in a rage.

Back then, at the beginning of the seventeenth century, calculating with numbers was not as commonplace as it is today. Most people still had the old Roman numerals in mind when they had to work something out. Due to their ignorance concerning numbers, many became easy victims for de Méziriac's think-of-a-number game. For it is obvious that the person who begins the game is guaranteed to win. The strategy is to begin by saying the number 1 and then follow with the numbers 12, 23, 34, 45, 56, 67, 78 and 89. This is always possible, no matter which numbers the opponent calls out in between. Of course, a crafty conman will never just roll out the number series 1, 12, 23, 34, 45, 56, 67, 78, 89, 100 like clockwork, but will act as if he is interested in the numbers said by his opponent. The game only poses a risk if the opponent begins and is thus the first to say a number between 1 and 10. However, as soon as he says a number that is different from those favored by de Méziriac—which is in all probability to be expected with an unsuspecting player—the conman's victory is in the bag.

In reality, the cunning Bachet de Méziriac was no conman. Rather, he had introduced his game as one of many diversions and mathematical puzzles in a book that he titled *Problèmes plaisants et délectables, qui se font par les nombres* (Pleasant and delectable problems, which are made by numbers). There was a highly receptive audience for the book in France at the time of its publication, since the French bourgeoisie and aristocracy were doing very well indeed. After the reign of Henri IV, the country gained great wealth and secure borders under first Louis XIII and then the regency of Cardinals Richelieu and Mazarin. Those who were not forced to scrape together a meagre existence as peasants or lowly laborers had few urgent or time-consuming tasks to fulfil and had plenty of opportunity to indulge in leisure pursuits. Having said that, very few of them knew what to do with this freedom, as Blaise Pascal perceptively noted: "Nothing is so unbearable to man as to be completely at rest, without passions, without business, without diversion, without occupation. He then feels his nothingness, his forlornness, his insufficiency, his dependence, his weakness, his emptiness. From the depth of his heart, uncontrollably will arise weariness, gloom, sadness, fretfulness, resignation, despair."[1]

Games offered release from this. Games enable us to while away the time and allow us to forget reality. They follow their own rules, to which we agree at the beginning and to which we then adhere, come what may. We know exactly when the game begins and when it ends.

Games create their own little world, where we can forget the vast, incomprehensible, and forbidding world we otherwise inhabit. They are indeed pleasant and delectable, and this is precisely what Bachet de Méziriac promises in his book.

Let us spend a few moments taking a closer look at his think-of-a-number game, this very simple con game. Like every other con game, it is in reality not a game in the proper sense of the word, since the conman does not play it for pleasure. For him, the con game is a kind of work. It may be disreputable, but it does provide a secure income. He himself will never experience the sheer joy of the game: at most, he will take fiendish delight in the opportunity to exploit his naive opponents. Conmen are not pleasant people and nobody with any sense should get involved with them. Unless one is from the police.

For the victim of the con game, there is nothing pleasant or delectable about it either. With brutal certainty, he is bound to lose in the end. The most malicious conmen temporarily allow their victims to win a little in order to keep them interested, only then to exploit them even more mercilessly—just as a cat "plays" with a mouse before administering the fatal bite, by seemingly opening up an avenue of escape, which then always turns out to be a trap.

But if we put aside such moral condemnation, con games are worthy of consideration, since they help us to gain a greater understanding of what the word "certainty" means. In the ruthless world of con games, it is certain who the victor is from the outset. There is no room for chance.

At this point, we can look back at Hans Hahn's lecture that had exerted such fascination on Karl Menger. On the left-hand side of the board, Menger sees the drawings of the smooth curves: the treble clef, the straight line to which the semicircular arch is joined. On the right, Hahn has drawn the erratic zigzag line so thickly that he has used up almost his whole chalk. On the zigzag line, the archetype of all share price performance graphs for listed companies, chance reigns. On the straight line, which contains no angles, one can at least approximately guess how Hahn will continue to draw it when one sees how he has sketched it so far. Admittedly, nobody could foresee that the semicircular arch would be added to the straight line. But the first few centimeters of this arch are hardly different from the straight line drawn before. At short intervals, at least, the continuation can be predicted for the drawing of a smooth curve.

Just as it is possible to predict how a straight line will continue, so too can one foresee how the rounds will end in bridge, a card game that demands great skill, if two experts are playing as a team against two newcomers who are veritable greenhorns in the game. Even if the beginners improve from round to round, this means only, to use a geometrical image, that the straight line representing their chances of winning rises imperceptibly above the zero level. We can draw an analogy between games in which the outcome of each round can be predicted with a degree of probability bordering on certainty and a completely straight line.

But which games correspond to the smooth curves? In the long term, we do not know where they will end. It is, however, always possible to guess accurately where a smooth curve will lead in the next few moments of its being drawn. If one imagines the tangent touching the curve at the current endpoint, the continuation of the curve will not depart too much from the tangent, at least not to begin with. The tangent represents the current trend in the game. Viewed in this way, smooth curves correspond to those games where one can predict the outcome of the next round, if not with complete certainty, then at least with a high degree of probability.

We can cite numerous examples of such games: in chess, for example, it quickly becomes evident which of the two players is the skillful expert and which is the absolute beginner. The physicist and statistician Arpad Emrick Elo, who was born in Egyházaskesző in Hungary but lived in the United States from the age of ten, invented a rating system in the middle of the twentieth century that gave chess players a so-called Elo rating that reflected their ability, using statistical evaluation of their previous tournament results. A player's Elo rating provides a relatively certain prediction of his or her chances of success against other players. Elo was astounded at the quasi-religious zeal with which people believed in the ratings given according to his system, saying, "Sometimes, I think I have created Frankenstein's monster. The young players are more interested in their Elo rating than in the pieces on the board."

Similar ranking lists are well known in sport. In tennis, there are separate world ranking lists for men and women, published by the respective tennis associations. For these, all of the points gained at the relevant tennis tournaments are added up and the player's ranking is thus determined. In soccer, the FIFA governing body publishes a world ranking list for national soccer teams. The World Football [Soccer] Elo Ratings, based on the Elo rating system in chess, are an alternative.

As Metin Tolan explains in his fine book *Manchmal gewinnt der Bessere: Die Physik des Fussballspiels* (Sometimes the better team wins: The physics of soccer), how certain the predictions are depends heavily on the game, or in this specific case the sport that one is observing. In tennis, you can rely on the previous strengths of the players more than in soccer, with one of the many reasons for this being that the tennis player remains the same person, whereas the people playing in the soccer teams are constantly changing. But even with handball or basketball, you can trust predictions more than with soccer. This is linked to the fact that there are relatively few goals in an average soccer game—generally speaking, two or three goals can be expected—whereas in handball or basketball, many more goals or points are scored. With a low number of goals, chance can quite easily put a wrench in the works of any predictions based on the relative strengths of the teams. It often happens that the clearly weaker team scores a lucky goal just before the final whistle and wins 2–1.

To use once again the analogy between the results of games and curves: games where the outcome is predetermined with a probability bordering on certainty correspond to the completely straight line. With con games, where the conman gives his victim absolutely no chance, a single point can replace the straight line. Games with a relatively predictable outcome, like chess or tennis when the players are ill-matched, correspond to the extremely smooth and slightly curved line. Games where chance can sometimes thwart the ability of the players are reflected in smooth curves that feature pronounced curvature like serpentines.

Of course, one mustn't take this rough analogy too seriously. But if we do take it further, the erratic zigzag lines like the "curve" scribbled on the right-hand side of the board by Hans Hahn represent the true games of chance, where luck alone holds sway and the outcome remains uncertain every time.

It is significant that the history of natural sciences can also be incorporated into this analogy. At least it can be when we vaguely, as though viewed from a distance, separate this history into three phases. To start with, there was the conviction that nature follows simple and unshakable laws, principles of complete clarity. The pre-Socratic thinkers Thales, Anaximenes, Parmenides, and their followers were convinced of this. Aristotle believed that "natural" motion in the terrestrial region, underneath the sphere of the moon, followed straight lines. These were the straight lines directed from above to below of the heavy objects consisting of earth or water, and those directed from below to above of the lightweight objects consisting of fire or air. From the moon sphere and far into the depths of the universe, there were only the perfectly circular paths of the heavenly bodies. Viewed figuratively, Aristotle's thinking on natural sciences was shaped by straight lines in the terrestrial region and circles in the heavenly sphere—the simplest curves that can be imagined.

The revolution of this philosophy came not via Copernicus, as is often claimed, but rather thanks to Johannes Kepler more than fifty years after Copernicus. Copernicus adhered to the idea of circular heavenly paths, whereas Kepler's calculations showed him that the planets' orbits were curves other than circles: "The orbit of the planet Mars is an oval," he wrote, referring to a curve with a constantly changing curvature. From 1609 onward, the year when Kepler's *Astronomia nova* (New astronomy) was published, smooth curves with their inconceivable diversity entered the world of physics and natural sciences. And fifty years after Kepler, the mathematics of differential calculus was invented, getting to grips with such smooth curves through the calculation of the tangents at all their points. In the simplest cases, this applied for all time. But such cases do not truly exist in nature. In the complicated cases that are reconstructed in line with nature, it is nevertheless possible to make very reliable predictions—sometimes, above all in astronomy, for centuries; sometimes, for example with the weather, for just a few days. And these predictions can be relied on to the extent that we can use them to shape the world: this is the basis for the technology and art of the engineers. Well into the twentieth century, our view of the world was influenced solely by smooth curves.

When Ludwig Boltzmann invented statistical physics in Vienna in 1900, and above all when Niels Bohr's ideas about quantum phenomena broke new ground, these previously smooth curves frayed into zigzag lines governed by pure chance. Only when viewed from afar do the curves still appear smooth—similar to when one roughly sketches the share price performance of a company over a long period, without taking into account each individual peak and trough, when one is purely interested in the long-term development. But in reality, pure and erratic chance rules within the world. Anton Zeilinger, the current grand seigneur of physics in Vienna and not dissimilar to Boltzmann in appearance, continuously points this out in his bestselling book *Dance of the Photons: From Einstein to Quantum Teleportation.* And he quite deliberately puts Einstein's name in the book title, for Einstein was the last physicist to believe that "God does not play dice"—and was completely wrong in this. At its core, physics is permeated with these zigzag lines.

Einstein's quote about God and dice was meant only figuratively, but his choice of words was not a bad one. Johannes Kepler was being much more serious when he wrote, following the ideas in his book about the harmony of the world: "We can see here how God, like a human architect, approached the foundation of the world in accordance with order and rule."[2] If we engage with Kepler's idea of the "architect of the world," the words just cited will lead us to envisage this architect as a creator, launching his world like a player. If nature were as Aristotle imagined it to be, the architect of the world would be an all-powerful player in whose hands the fabric of creation flows like molten wax. Then, the outcome of the game—that is to say, the manner of all things for all eternity—would be unshakably determined. Everything would be predestined from the beginning, like fate, and what happens in nature would be dreadfully dull. Now, after Kepler and after Boltzmann and Bohr, we believe that the hands that nature plays proceed differently—with myriad variations, though broadly predictable. The architect of the world is engaged in a game consisting of a countless number of rounds in which one can at least partly sense the trend of how the game will proceed. Although Einstein maintained the opposite view until the end of his life, it remains a fact that he was wrong—the architect of the world does not have full control over his creation. He seems to throw a dice for every single round and let pure chance take its course on the arena of the universe.

If we wished to continue this idea, numerous questions would immediately arise: What profit does the architect of the world gain from his game consisting of such an immense number of rounds? Who are his fellow players? Are they the "intelligences hidden in the clusters of star dust and separated by billions of parsecs" about which Stanisław Lem writes in his book *A Perfect Vacuum*? Or is it actually we, every single one of us? Do the other players lose when he wins, or do they win along with him? What exactly does the word "win" mean in this context? And how do we know when the game has reached its end?

It must have been questions like these that Goethe was wrestling with when he wrote *Faust*. Faust wishes to master the game of the world and Mephistopheles is having his fun with Faust, but both of them are actually the plaything of some other force. Goethe's *Faust* is a work of art and not a scientific treatise on the questions cited—to hope otherwise would be foolish. All of these analogies between games, curves, and the creation of the world are only meant symbolically, and in symbolic speech there are no definitive answers to child-like questions.

And so we return to tangible reality in order to investigate what the term "chance" actually means and what role it takes in games.

4

PLAYING WITH CHANCE

PORT ROYAL DES CHAMPS, NEAR PARIS, 1655

"Monsieur Pascal, I have a problem."

Antoine Gombaud feels ill at ease in the cell where Blaise Pascal is staying. Everything about Port Royal des Champs makes Gombaud uncomfortable. What on earth is he doing in this convent near Paris that is home to peculiar characters—nuns who scurry along the corridors with a constant pallid smile on their faces? He has no time for the ideal of holiness to which the nuns are devoted and which is all about the study of inner emotions and impulses, praying, fasting, poverty, self-torment, and works of charity. And yet it is precisely here where Pascal, once so gregarious and ready for pleasures of all kinds, has retired to spend his days in a room by the convent walls that has been specially prepared for him, grave and introverted and thinking only of his salvation. Gombaud is convinced that his friend has changed completely.

Gone are the times when they would talk long into the night about anything and everything, about God and the world—especially the world with all its desirable aspects; when they would roam through Paris from one pleasure spot to the next, leaving no salon unvisited and succumbing to earthly delights—he, Gombaud, happy with the more banal kind as well, but Pascal only interested in those that also gave him intellectual enjoyment; when they would remain at the gambling tables until dawn and indulge in games of dice and cards.

Gombaud, a wealthy man, was anything but coarse and unrefined, despite his addiction to pleasure. On the contrary—he was a writer and was even considered to be an expert in matters of courtesy and *savoir faire.* Under the pseudonym of "Chevalier de Méré," he wrote dialogues about the nature of the true aristocracy and the character of a noble man. His friends addressed him by this pseudonym henceforth and, even today, it is the name by which he is commonly known. The height of his ambition, however, was to win at games of chance. He devoted himself incessantly to them, particularly now that the most intelligent of his companions, the ingenious Blaise Pascal, had left him and withdrawn into reclusive solitude.

The problem that now occupies Gombaud, however, will not leave him in peace, and he feels compelled to speak to Pascal about it. For Pascal, the cleverest man in France at the time of Richelieu and Mazarin, is probably the only one who knows the answer. And so Gombaud takes a carriage to Port Royal des Champs, knocks at the door of the spartan cell and sees Pascal for the first time in a long while, austere in appearance and emanating an aura of holy solemnity.

"My dear Pascal, there is a problem that has been bothering me for a long time," begins Gombaud again. "You know that I am partial to gambling." He looks into the eyes of his former companion, who is imperturbably scrutinizing him and seems to be concerned with quite different, weightier problems, and Gombaud realizes how trivial the matter in question is. Nevertheless, he plucks up his courage and continues: "On Thursday last week, I spent the whole night at the games table with a friend, it doesn't matter who it was, and just before dawn, we agreed to proceed like this for the last game: whichever one of us would be the first to win four rounds would win the entire stake. We had fixed the stake at 60,000 livres."

"Naturally. Games need a certain thrill."

"Quite right." Gombaud hesitates slightly and looks around the room, which contains nothing that might cost even 10 livres, before continuing: "To begin with, luck smiled upon me. I won three times in a row. Then my friend won two rounds. And then it happened—with dawn already breaking, and before we could prepare the dice for the next round, one of the king's couriers charged into the salon and commanded my friend to ride with him immediately to the royal palace. That was of course a *force majeure*, a higher force that compelled us to stop the game. So now there we are, my friend and I, and we don't know how to divide up the 60,000 livres fairly between us."

"And you have come to me with this question," says Pascal quietly, and Gombaud isn't sure whether he is simply making a sober statement or if a faint accusation can be discerned.

In any case, Gombaud immediately feels the need to justify the question. "Naturally, I researched the matter before coming. As early as the end of the fifteenth century, problems like this were being pored over in Italy. Luca Pacioli . . ."

"The inventor of double-entry accounting; a very clever man," interjects Pascal.

". . . this same Luca Pacioli suggested dividing up the stake according to who had won the previous rounds. This would mean that I would receive three fifths of the stake for my three wins and my friend would get two fifths for his two wins—36,000 livres for me and 24,000 for him. But our mutual friend Mersenne, whom I also told about this, referred me to *The Book on Games of Chance* by Gerolamo Cardano. I read Cardano's writings and came to the conclusion that he contradicts Pacioli. Cardano is of the opinion that the prior course of the game is unimportant; what matters are the possibilities for how the game can end. And I have been through all of these possibilities. First, there is the possibility that I win the next round. Second, there is the possibility that my friend wins the next round and I win the one after that. And third, there is the possibility that my friend wins the next two rounds. The first two possibilities are favorable for me, while the third one suits my friend. If we look at it in this way, I should receive two thirds of the stake—i.e., 40,000 livres—and my friend should receive one third of the stake—i.e., 20,000 livres. I don't want to spoil my amicable relationship with my friend, however, which is why I have come to you, Monsieur Pascal, to ask which of the two, Pacioli or Cardano, is right."

"It's an interesting question," says Pascal, and Gombaud is relieved that he is no longer so taciturn. "I would lean toward Cardano's solution, but I'd like to reflect on it a little. Can you give me time for that?"

"As long as you like," Gombaud hurriedly assures him.

"Allow me to ask two or three more questions. You could simply arrange with your friend to forget the rounds already played and each of you gets back your stake of 30,000 livres. How about that?"

"That's out of the question," sputters Gombaud. "Once the game has begun, you can't take anything back. I could just as easily claim to my friend that the previous rounds had shown that Lady Luck obviously favored me more than him and so I should claim the entire stake of 60,000 livres. But that would be equally unreasonable, in my view. No, my friend and I would like a fair division."

"I understand. And the chances of winning each round were the same for both players? It was a pure game of chance, not like chess or a game of skill?"

"I can assure you: in every round, each of us had the same chance of winning."

"And you are absolutely sure that your friend is not a conman, luring you into a false sense of security by losing three rounds, before winning four times in a row and pocketing the entire stake?"

"Monsieur Pascal!" says Gombaud indignantly, and now it is the Chevalier de Méré who is speaking through him, "We are men of honor!"

"Ah yes," says Pascal under his breath, and you can faintly hear in his tone the memory of bygone times. He holds out his hand to Gombaud: "I will write to you."

After the door has clicked shut, Pascal remains sunk in thought. He is convinced that Cardano's approach of considering the possible future scenarios could be helpful, but he isn't happy with the way that Gombaud has listed the possibilities.

What possibilities are there for the next round? Clearly, Gombaud could win (Pascal writes the letter G) or his friend could (Pascal writes the letter F beneath). What possibilities are there if two rounds follow those already played? Either Gombaud wins both rounds (Pascal writes GG). Or he wins the first and his friend wins the second, or the other way round (Pascal notes these two possibilities as GF and FG beneath). Or Gombaud's friend wins both rounds (Pascal writes FF beneath). If we take a look at Pascal's sheet of paper, we can see the following combinations of letters:

G	GG
F	GF
	FG
	FF

Now Pascal considers: if we take into account the rounds already played, Gombaud's friend will only succeed in being the first to win four rounds in the last scenario, FF. In the other three scenarios, Gombaud would be the victor, since he has already won three rounds. Accordingly, the 60,000 livres should be divided up with a ratio of three to one: 45,000 livres for Gombaud and 15,000 for his friend.

But Pascal is nagged by the doubt that the second possibility, GF, that he has written in the second column cannot come to pass; should Gombaud win the next round, the game is over. This would require the following correction to the second column of Pascal's notes:

G	G~~G~~
F	~~GF~~
	FG
	FF

Had Cardano perhaps hit the mark? Pascal refuses to believe it. Deep down, he senses that he is on the right track with the ratio three to one and that his doubts about it can be dispelled. But he cannot yet see how to undo the correction just made.

Among the many great intellects living in France at that time, one particularly stood out in the field of mathematics: the lawyer Pierre de Fermat, who worked in Bordeaux and Toulouse. He was well known in the Parisian academic world, even though he never spent any time in Paris. The reports that he wrote in letters to members of the academy led by Mersenne—Pascal had been a member of this academy since his early years—gave him renown as a strange, yet ingenious amateur mathematician. He constantly came up with odd claims, mostly concerning the mysterious prime numbers, about which nobody knew the rules according to which they followed on from one another. Fermat never revealed the proof behind these claims. He simply invited his correspondents and their colleagues to substantiate his claim. This would give them the opportunity to measure themselves against his talent. Nobody really succeeded in this at the time, but none of Fermat's contemporaries could demonstrate that he was mistaken in his claims, either.

Pascal therefore decides to write a letter to Fermat about the *force majeure* problem facing Antoine Gombaud, otherwise known as the Chevalier de Méré, before he composes his reply to Gombaud himself. Quicker than expected, he receives a response from Toulouse, in which Fermat confirms that Pascal is right to suggest a ratio of three to one—i.e., 45,000 for the Chevalier de Méré and 15,000 for his friend. A further exchange of letters between the two scholars develops, in the course of which they expose the word *probability* as a notion that can be quantified with fractions between zero and one.

The probability that Gombaud wins the next round is one-half, or 50 percent. If he does, the game is over and Gombaud has won the stake. The probability that another round will be played is also 50 percent, and with half of this probability—i.e., with a probability of 25 percent—Gombaud's friend will win that round. Thus, the probability that Gombaud's friend gets the stake is only 25 percent. Even if there is another round, Gombaud can claim the other 25 percent of the probability of winning the stake. All in all, therefore, the probability that Gombaud is the first to win four rounds is 50 + 25 percent, that is to say, 75 percent.

This means that the correction on Pascal's sheet of paper needs to be revised. Pascal fills in the probability of each individual scenario and adds up the values noted by G and FG, the outcomes that are favorable for Gombaud:

G	GG	50% for Gombaud
F	GF	
	FG	25% for Gombaud
	FF	25% for Gombaud's friend

Fermat illustrates the same idea with the picture of a tree growing upward. The thickness of the trunk protruding from the ground is one, since the probability that the game must go on is one—i.e., it is 100 percent certain. The trunk then splits into two branches, the thickness of each of these being one-half. One branch leads to the scenario that Gombaud wins the next round, where the tree ceases to grow, since the game is over in this case. The other branch leads to the scenario that Gombaud's friend wins. It then splits into two smaller branches, each with a thickness of ¼. One of these branches ends in the scenario that Gombaud wins and therefore receives the stake. This, added together with the probability of one-half at the end of the previous branch, means that Gombaud has a total probability of three-quarters, or 75 percent that he receives the stake. It is only at the end of the other smaller branch that the possibility materializes that Gombaud's friend receives the stake. The probability of this is one-quarter, or 25 percent.

What is particularly pleasing about this solution to the Chevalier de Méré's question is that the *force majeure* problem can be solved even in cases where the chances of winning or losing a round are not fifty-fifty. Let us assume that Gombaud and his friend are playing chess and the many games that they have played against each other in the countless nights before show that Gombaud is much the better player. On average, he beats his friend four times out of five, losing only once. Accordingly, the probability that Gombaud wins the next round is 80 percent. The two branches of the tree are now of different thicknesses, with the one leading to Gombaud taking up four-fifths of the trunk, while the one leading to his friend only takes up one-fifth. And only this thin branch is split once again, into two smaller branches, of which the one leading to Gombaud is four times as thick as the one leading to his friend. 20 percent of 20 percent, that is, just one-fifth of 20 percent—i.e., a mere 4 percent—that is now the probability that Gombaud's friend will win that particular game too. Therefore, his fair share of the 60,000 livres would be only 4 percent—i.e., 2,400 livres—while Antoine Gombaud can claim the remaining 57,600 livres.

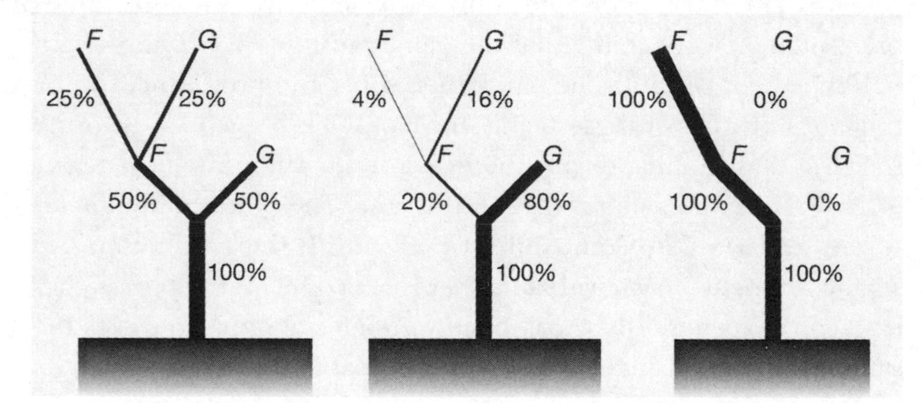

Figure 4.1: The three tree diagrams by Pierre de Fermat show the possible ways in which the game between Gombaud (G) and his friend (F) could proceed. In the left-hand diagram, the chances of winning each round are fifty-fifty. In the middle diagram, Gombaud wins four out of five rounds on average. In the right-hand diagram, Gombaud is being tricked by a conman and so has no chance of winning.

Even if Gombaud is mistaken about his supposed friend and has indeed been taken in by a conman, the probability calculation still provides the correct answer for the *force majeure* problem. In this case, the probability that Gombaud wins the next rounds is absolutely nonexistent. At the first fork in the trunk, the branch leading to Gombaud is dead and the other branch leads to the next fork with the same thickness as the trunk. There again, the branch leading to Gombaud shrivels and dies before it even starts, while the other branch, still just as thick as the trunk, leads to Gombaud's malicious "friend," who ruthlessly pockets the full 100 percent of the stake.

This is the harsh reality of con games and, in such a case, Gombaud could count himself lucky that the king's courier's sudden entry into the salon brings about a premature abandonment of the game.

Many people believe that the probability theory devised by Pascal and Fermat can open a window into the future. This is indeed correct. But we must never forget that the glass of this window is translucent, allowing not a clear picture but rather one that is dull and opaque.

We can most easily see how this is to be understood by using dice. The symmetry of the die enables us to establish beforehand how great the probability of throwing a six is: each of the die's sides is indistinguishable from the others, if we discount the different numbers of spots engraved or drawn on each side. With a "fair die," therefore, the probability of throwing a six is exactly one-sixth, or just under 16.7 percent. Only conmen use "unfair," so-called loaded dice. They might fit a piece of lead into the die near the side with one spot, for example. Number one is on the opposite side to number six—the spots on two opposite sides always make up seven in total—and so the weight of the lead will cause the dice to land on six when thrown. More brazen cheats use dice that have six on two sides, not only one. This means that the probability of throwing a six is doubled, from just under 16.7 percent to just over 33.3 percent.

As honest players, however, we use a fair die. We know that the probability of throwing a six is about 16.7 percent. For our next throw, this knowledge is of literally no use. It does not help us in the least. We might throw a six, or we might throw another number—the probability of each number is the same.

If we decide to throw the die twelve times, then we can make a bit more use of our knowledge about the probability of throwing a six. We would be surprised if a six did not appear a single time in those twelve throws, although it is by no means impossible. But we have good reason to expect two sixes—regardless with which of the twelve throws—and would not be surprised if we threw a six once, or even three times. It would be odd, however, if we threw a six all twelve times. In such a case, it would be wise to check carefully whether the die is indeed fair.

We can only make truly reliable statements if we throw the die a great number of times. If we throw a fair die six thousand times, we can expect that the number six will appear about a thousand times—not exactly, but not far off. This is called the "law of large numbers." Pascal and Fermat must surely have anticipated this, but it was Jakob Bernoulli, along with his brother Johann a forebear of an extensive dynasty of Swiss scholars, who first put it into words. It appeared in his work *Ars Conjectandi*, or *The Art of Conjecturing*, which was published by his nephew Nikolaus Bernoulli in 1713, eight years after Jakob's death.

Probability does indeed open a window that, though it may be made of translucent glass, does allow a view of the future. If a die is thrown 12,000 times, we can bank on about 2,000 sixes. For throw number 12,001, however, following the first 12,000 throws with around 2,000 sixes, we still do not know which number will be shown, despite the concept of probability. Our view of the isolated event remains clouded.

Nevertheless, it is possible to use our knowledge of probability to our profit—in the truest sense of the word.

5
PLAYING WITH TIME

PHILADELPHIA, 1746; AMSTERDAM, 1636–1637

"Time is money."

These famous words originate from Benjamin Franklin, one of the founding fathers of the United States of America. Franklin was a true polymath. Born in Boston in 1706 as the son of a soap and candle maker, he first worked as a printer and publisher, before retiring from business at the age of forty-two and going into politics. He was active as a writer, scientist, inventor, and statesman, and he established the first volunteer fire department in Philadelphia and the first public library in America, while his inventions included a particularly effective and smoke-free stove, the lightning rod, the glass harmonica, the flexible urinary catheter, swim fins, and bifocal glasses. He had a key role in drafting the United States Declaration of Independence, successfully represented the newly founded country as the ambassador to France, and was instrumental in developing the United States Constitution. Even at the age of eighty-four, in the final year of his life, he was campaigning publicly for the abolition of slavery.

A customer enters Franklin's book and newspaper shop in Philadelphia. He has his eye on a certain book on the shelves and asks the young shop assistant how much it is.

"One dollar," is the trainee bookseller's short answer.

The customer tries to negotiate. "One dollar—that's quite a lot. Is there a discount for cash?"

"Not that I know of. Our prices are fixed."

"Well, some sort of reduction must be possible. Let me speak with Mr. Franklin himself."

The young man hurries into the neighboring newspaper editorial office and interrupts Franklin, who is in the process of editing the current issue: "A gentleman in the shop would like to speak to you, sir."

"Not now. I'm busy, as you can see."

"But I don't know what to tell him, sir. He wants the new book, but one dollar is too much for him. And I am not allowed to negotiate on price—you forbade me to, sir."

Full of annoyance, Franklin throws his pencil to one side, strides into the shop, and growls at the customer, who is initially pleased to see him approaching: "We charge one and a quarter dollars for the book."

"But your salesman said it was just one dollar," replies the customer in bewilderment.

"You should have taken it for that price! Instead, you're keeping me from my work."

"I will take your words as spoken in a humorous vein, Mr. Franklin," the customer says in a pacifying tone. "Now, let's be serious: one dollar is, in my view, too considerable a sum. What is the lowest price that you can offer me?"

Now Franklin becomes truly angry. "One and a half dollars! And the longer you take up my time, the more expensive it will become!"

We do not know if a deal was ever made. But this anecdote illustrates Franklin's understanding of his saying "Time is money," which appeared in his pamphlet *Advice to Young Tradesmen*: we should use our time for valuable work that helps us earn an honest living.

It is to be welcomed that a large number of people are able to escape poverty thanks to industrious work, either with their hands or minds. But nobody will ever become genuinely wealthy through zealous hard work. In order to become truly rich, it is necessary to understand the phrase "Time is money" in a much broader sense than the self-made Franklin meant, namely in the following way, as recounted by the successful Viennese entrepreneur and businessman Michael Gröller: When a bold young man once asked Louis Nathaniel de Rothschild, the last of his dynasty in Austria, how his family had attained their immense wealth, Rothschild is said to have answered with a roguish grin: "I'd be glad to tell you. We always sold a tiny bit too early."

Rothschild thus reveals the secret of how to become rich: the key is to be able to see into the future, which is how one can then choose the right moment to do business successfully. Among the twelve sons of Jacob, the patriarch of the Israelites, only Joseph, the cleverest of them all, had the gift of being able to see clearly into the future and to know that the seven years of plenty and rich harvests would be followed by seven years of famine. This meant he was in a position to advise the Pharaoh to store the grain harvested in the good years and not to sell it immediately. For if supplies are plentiful, demand is low, and the price for a sack of grain is next to nothing. It is only in the lean years that it pays to take the grain out of the granaries and sell it to the hungry people of the neighboring lands, when everyone is starving for the corn that is now worth its weight in gold.

We do not have Joseph's gift, however, and our dreams do not reveal the future. We therefore have to rely on probability theory, which tells us to what extent we can count on certain events happening in the future. We can best understand how to use this to our profit by using the game of dice as an example—a very simple game, admittedly, not to say crude, but one that makes the matter easily apparent.

To be successful, the game requires a large number of compulsive gamblers, all of whom come to the organizer of the game. He has a die in his hand and tries to tempt each passerby to throw the die: "I'll give you four hundred dollars if you throw a six," he promises each one.

But before the player receives the die from the organizer, the latter demands a stake of one hundred dollars. "You'll get the stake back if you throw a six," he tells the player reassuringly.

"And if I don't throw a six?"

"Then you lose the stake," the organizer admits sadly, but his face brightens immediately: "But just imagine—if you throw a six, you get your stake back and, on top of that, you win four times as much, so four hundred dollars."

If we assume that six thousand compulsive gamblers take up the offer, then the organizer has a total pot of 600,000 dollars, since each player has pressed one hundred dollars into his hand. Of course, the organizer has to pay five hundred dollars out of this pot to those lucky enough to throw a six. He is happy to do this, since he can use probability to calculate that he has chance on his side. If the dice is thrown six thousand times, the law of large numbers dictates that the number six will come about a thousand times. He will therefore hand out five hundred dollars to about one thousand lucky winners, making a total of about 500,000 dollars. As a result, he is left with about 100,000 dollars in pure profit.

This profit is a mathematical certainty. The law of large numbers can be relied upon, and it is just as certain as the fact that six times seven is forty-two.

It would also be wrong to dismiss all the players who enter into this pact as stupid. They are only stupid if the loss of their hundred-dollar stake causes them anguish, in contrast to those who are so well-off that having a hundred dollars more or less in their pockets makes little difference to them—we can think back here to the fifth sack of wheat in Carl Menger's idea of the clever farmer. On the other hand, people who don't care about losing a hundred dollars are also rich enough for the profit of four hundred dollars to make no difference to them, either. What would tempt them to play this game would be at most the allure and thrill of uncertainty and the suspense of wondering whether luck is smiling upon them or not.

For every single player, the future—limited here to the uncertainty of whether the die will land on six or not—is completely unknown beyond the 16.7 percent chance of winning. For the organizer, however, if he can convince six thousand fortune-hunters to throw the die, his future winnings are guaranteed.

The question immediately arises, of course, as to why there isn't some dubious chap on every street corner, proposing the game of dice just described to passersby.

The answer can be found in the game's banal and crude nature. Anybody for whom a hundred dollars is a significant sum would refuse to play—assuming he isn't completely stupid—because of the excessive risk involved. Rich people, on the other hand, are only interested in winning much more than four hundred dollars, and most of them wouldn't want to waste their time with such a trivial and simplistic game.

Nevertheless, since time immemorial, this game has been successfully practiced in a variety of guises and forms.

"Darling, could you join us, please—the insurance man is here!" calls the anxious husband to his wife, for it is time to conclude some important business.

"You've certainly bought yourselves a magnificent house," says the insurance agent admiringly, as he sits down in the proffered armchair in the elegant living room and takes his papers out of his briefcase.

The house owner and his wife nod, pleased. "We're so glad to live here," she beams. "It certainly cost us a pretty penny, and without the inheritance that my husband received, we wouldn't have been able to afford such a beautiful mansion."

"That's why we want to take out fire insurance," adds her husband.

"Well, you're in good hands with us," says the agent reassuringly. "Our company arranges stand-alone fire insurance for private customers, too. I have already prepared the contract in accordance with your wishes, as already discussed in our office. We will insure your house so that, should it be destroyed—God forbid!—by a fire, an explosion, a plane crash, or a lightning strike, we will provide financial compensation for the damage. Your beautiful house will be insured for two million dollars."

"Then let's sign the contract straightaway," says the husband immediately, but his wife hesitates a moment. "How much is the monthly premium we have to pay?" she asks.

"It's barely worthy of mention," says the insurance agent jovially. "The annual premium is just 468 dollars, so that's a mere 39 dollars per month."

"And why that particular amount?" The wife is not as easily convinced as her husband.

The agent attempts to appease her with an evasive smokescreen. "You're welcome to contact our competitors. Nobody in our industry offers better conditions for fire insurance than we do."

This deflects attention away from the actual answer to the wife's question, which is that the transaction being concluded here is nothing other than a variation of the game of dice described above. The annual premium of 468 dollars is the "stake" that the "player," in this case the insured party, pays in order to be allowed to participate in the "game"—i.e., to be able to exercise his rights by signing the insurance contract. Throwing the die is the equivalent here of observing what happens to the house over the year: if a fire destroys the house, this is analogous to throwing a six, because in that case, grotesque though this may sound, the insured party has "won the game," since he receives two million dollars as "winnings." On the other hand, the probability that the house burns down is not 16.7 percent, like that of throwing a six in the game of dice, but is in fact considerably lower, indeed virtually nonexistent, since it is very rare for houses to be destroyed by fire. In its decades of observations, the insurance company records how many houses in a million are destroyed by fire, lightning, an explosion, or a plane crash over the course of a year. For the sake of simplicity, let us assume that this is true of one hundred houses. This means that there is a probability of one-hundredth of a percent that a randomly selected house will burn down during the year.

In order to keep the calculations simple—we are after all only concerned with the principle here—let us assume that the agent's insurance company has insured 100,000 houses, each for the sum of two million dollars. The company gets 468 dollars from each of the 100,000 house owners, which gives it an income of 46.8 million dollars per year. The company must, however, factor in the probability that, within the year, 0.01 percent of the 100,000 houses—i.e., 10 houses—will be completely destroyed by fire, lightning, an explosion, or a plane crash and thus ten times two million—i.e., a total of 20 million dollars—will have be paid out as compensation. The company is happy to pay this money, because it still retains 26.8 million dollars each year, which can be put to good use for government taxes, staff wages, investment in fire protection measures, lobbying activities, and above all for reserve funds. Plus for paying private detectives to find out whether a fire has been caused by arson carried out by the insured party or some rogue employed by them, for in such a case the insurance company is not required to pay out, because the insured party has cheated. By way of analogy with the game of dice, he has paid the hundred-dollar stake to the game's organizer, taken the die, found the six, and then, rather than throwing the die as normal, carefully placed it so that the six is facing up. In this way, the cheat has taken the role of chance out of consideration but has also destroyed everything that actually makes up the game.

Despite the similarities in principle between the simple game of dice on the one hand and the insurance business—as just one example of the myriad varieties of such business models—on the other hand, there are nonetheless significant differences, of which we can focus on two in particular:

First, in the game of dice, the potential winnings of four hundred dollars are not especially attractive when compared to the stake of a hundred dollars. By contrast, compared to an annual insurance premium of 468 dollars, the potential "winnings" of two million dollars—insofar as such "winnings" can make up for the loss of a house—represent an attractive proposition. The insured party can say to himself that, if he pays about 500 dollars each year, it would take him four thousand years to pay the equivalent of his "winnings." The cost incurred by himself and his children and their children is, from his point of view, far less significant than the potential "winnings."

This is why, unlike our simplistic game of dice, which is undoubtedly of limited appeal, other games of chance such as roulette or the lottery are very popular among the general population. These games offer people the chance to take home substantial winnings for a relatively low stake, and this acts as a magnet to speculators. We will address this in detail in the next chapter, focusing in particular on the special pitfalls of roulette.

Second, the insurance agent's final words in our fictitious dialogue, when he challenges the lady to compare the competitors' offers, can be taken very seriously indeed. His answer may have evaded the actual question, but his words are nevertheless significant. The annual premium of 468 dollars to insure a house for two million is no more carved in stone than the probability of damage established by the insurance company, since, in most business scenarios, probabilities change over time.

Our example of fire insurance is a good illustration of this: a few decades ago, the likelihood of a house burning down was by no means insignificant. There were open fireplaces, flames on the stove, and even light was provided by candles and petroleum or gas lamps. There was therefore a constant risk of fire being caused by negligence or poor maintenance. Due to the higher frequency of fires at that time, the estimated probability that one of the houses insured by a company might catch fire during the year was much greater than today. Accordingly, the annual premiums were significantly higher.

This is why insurance companies have a burning interest in the improvement of fire protection systems and are constantly lobbying for fire protection regulations to be tightened, the aim being to keep the likelihood of a fire low and to reduce it in future still further. Because the lower the annual premium is, the more homeowners are prepared to take out the insurance, and since competing companies don't rest on their laurels, those who can provide the same service for a lower price are bound to hold the winning hand.

It is not out of love for science or because they don't know what to do with all their money that companies invest in research and development; rather, they hope that improvements to the technology already in use, or even technical inventions, will change the existing probability of making a profit with a product to their advantage. Of course, only a charlatan promises to guarantee a product's success. Serious business men and women are fully aware that they can only count on probabilities, which is why it is crucial for them to be able not only to count on these but also to predict their development. Things can end tragically when, based on a short-term trend, people bet prematurely on a lasting change to these probabilities.

To illustrate this point, we can go back to the year 1636 in Holland, where there was a veritable craze for tulips, which had been introduced not long before from the Ottoman Empire. Tulips were *the* flower in fashion at the time. Women were delighted by their colorful splendor and variety, and tulips were put everywhere—not only in every apartment room or in every vase in the salons, but also on clothes and every other place imaginable. When women want tulips, men have to buy them. The problem was that there were no more tulips in the shops; demand was too great. Only tulip bulbs could still be bought, for horrendous prices, and even these soon became scarce.

"I need five tulip bulbs, now!" cries a Dutch burgher, storming into a flower store.

"I'm sorry, Mijnheer Grebber," answers the store owner, "My next delivery isn't until February next year."

"I'll pay anything," says Jan Grebber impatiently, but the store owner suggests a deal.

"Mijnheer, we can make the following arrangement. Today, a Semper Augusta bulb, the most beautiful tulip that I sell, costs a hundred Gulden. I will sell you not only five, but ten of these bulbs, but I can only deliver them to you in mid-February, 1637."

"You want me to pay the huge sum of a thousand Gulden and will only deliver the bulbs in half a year's time?" splutters the customer, but the store owner quickly pacifies him:

"No, no. You don't need to pay for the bulbs now. Only on delivery. Just think—the way the price of tulip bulbs is going, a Semper Augusta bulb will cost five hundred Gulden at the beginning of next year. But I will stick to my suggested price of one hundred Gulden per bulb. You will receive ten bulbs in mid-February, payable on delivery. Then you will probably be in possession of a treasure worth five thousand Gulden and will only have to pay one thousand for it. Isn't that a lucrative deal?"

"It certainly sounds tempting," admits Jan Grebber.

"Of course, you must understand, Mijnheer Grebber, that I too must benefit in some small way from our agreement. I too would like to profit from it. If we agree on this deal, that you have the option to buy ten Semper Augusta bulbs for one hundred Gulden per bulb, to be delivered and paid for on February 15, 1637, then I ask a thousand Gulden for issuing this warrant."

"For the warrant only? But that is as much as ten tulip bulbs of the very best kind cost today!"

"For the warrant only and on immediate payment. These are the conditions of the deal."

Jan Grebber makes a rapid calculation. He only actually needs five bulbs for himself, so he can sell the other five bulbs after delivery in February. With a probable price of five hundred Gulden per bulb, he will get 2,500 Gulden for the five bulbs. Given that he has to pay the trader a thousand Gulden for the warrant now and then a thousand Gulden for the actual bulbs, he will be left with a probable profit of five hundred Gulden. He would get the five tulip bulbs, which he needs anyway, for free and, on top of that, get some money too. With these thoughts in mind, he agrees to the deal with the flower trader and, after the warrant has been issued, presses a thousand Gulden—a fortune at the time—into the trader's hand.

The only thing is that probability changes over time.

To begin with, it does so to the advantage of Jan Grebber, who is resplendent with joy when prices for tulip bulbs rocket in December 1636. It is not uncommon for the thousand Gulden mark to be exceeded for the price of a single Semper Augusta bulb. At the end of January of the following year, the prices are so high that Jan Grebber feels as though he is floating on air. But on February 3, at an auction in Haarlem, none of the tulips on offer are sold at the expected prices. In the following days' auctions, too, the traders either remain stuck with their bulbs or are forced to sell them at unusually low prices. It seems that the scales have fallen from the buyers' eyes, and they have realized that tulips have no lasting value and are nothing more than mere, useless flowers.

Jan Grebber feels queasy. If the price continues to collapse, and the likelihood of this happening is as great as that of the price rising in December, then his money is down the drain. On February 15, the due date of the deal, the trend of the tulip price graph proves to be accurate. The few upward spikes make no difference, and the downward spikes multiply at a fearful rate. It is now possible to buy a Semper Augusta bulb for a single Gulden. The thousand Gulden Jan Grebber paid for the warrant, enough for a property in Amsterdam, might just as well have been thrown out of the window. The change in the tulip price has made him a poor man.

"I want my thousand Gulden back! Here's your dirty little warrant!" the bankrupt burgher shouts at the flower trader.

"You can tear it up, as far as I'm concerned," answers the trader surlily. Quickly pressing a bag of ten tulip bulbs into Grebber's hand, he shows him to the door.

6

PLAYING WITH A SYSTEM

PARIS AND PORT ROYAL DES CHAMPS, 1659; ST. PETERSBURG, 1738

La Théorie de la Roulette.

A book with this title, a highly seductive one for gamblers, was published in 1659, written by the completely unknown Amos Dettonville. If you rearrange the letters of the author's name, however (U and V are considered to be the same letter in the old Latin alphabet), you get Louis de Montalte, the name of the author of *Lettres Provinciales,* a highly popular and well-known book in France at the time. It consisted of a collection of eighteen acerbic letters written by the secretive Louis de Montalte, criticizing the excesses of the Jesuit order in France. Both Louis de Montalte and Amos Dettonville are pseudonyms, used by France's greatest genius, Blaise Pascal, who is already familiar to us from his discussion with Antoine Gombaud, the Chevalier de Méré.

There is a temptation to translate *La Théorie de la Roulette* as "The Theory of Roulette." Unfortunately, such a literal rendering would be wrong. In his book, Pascal did not address this game of chance, which enjoyed immense popularity even then and which had its origins in the old "wheel of fortune." Instead, the book was all about so-called cycloids, which he called "roulettes" in French.

Nevertheless, Pascal is indeed said to have been the inventor of a method whereby one can be sure of winning in roulette, a system that enables one to use chance in such a way that monetary profit can be gained from it.

It is obvious that games of chance are particularly well suited for people to try to come up with winning systems. Unlike with other games, where the probability of winning depends to a large extent on the skill, experience, and sophistication of the players, as well as on a number of other external factors, the probabilities in games of chance are strictly defined, widely known, and not subject to change. In addition, games of chance do not involve the fickle risks that occur in business and are due to the fact that the probability of winning is dependent on time—we described in the last chapter the bitter experience that people can have with this.

In this respect, roulette is the simplest game imaginable. It is a variation on the basic game of dice seen in the last chapter, with the six-sided die replaced by a wheel spinning in a basin, with its thirty-seven numbered slots lining the outside of the wheel. The probability of one in six that a particular number is thrown with the die is replaced by the probability of one in thirty-seven that the ivory ball falls into a particular slot after being released. While in dice, winning means you receive five times the staked amount, if you include the stake, in roulette, the payout is thirty-six times the staked amount (including the stake), even though there are thirty-seven slots, cleverly numbered from zero to thirty-six by the game's inventors. The fact that players can bet not only on individual numbers, but also on various groups of numbers, such as only odd or even numbers, only black or red, or many other variations, is a major part of the game's appeal for many people. In principle, however, it is incredibly simple.

It goes without saying, therefore, that if there is a system that promises the chance to win at a game with absolute certainty, and this game is not a con game, then it must be with this archetype of all games.

"I'll bet 100 livres on red." During one of his visits to Pascal, Antoine Gombaud had learnt the strategy for the winning system developed by his friend. He tosses the colorful 100-livre chip to the croupier, who places it, as requested, on the felt-covered table in the section with the red diamond.

"Twenty-seven, passe, impair, rouge," calls the croupier, after the ivory ball has dropped into the slot marked twenty-seven. The French word "passe" shows that the number twenty-seven is greater than eighteen, as opposed to "manqué," which refers to numbers one to eighteen, while "impair" shows that twenty-seven is odd, as opposed to "pair" (even), and "rouge" shows that the slot marked twenty-seven is red and not black ("noir"). Gombaud, having bet on red, now receives double his stake—200 livres.

"If you win, you should go home," Pascal had advised him. But winning 100 livres—the equivalent of a craftsman's average annual salary—is not enough for Gombaud. As a passionate gambler, he feels compelled to start again from the beginning. "Start off with a small stake," Pascal had advised.

"100 livres on red once more," he says, and the croupier places Gombaud's chip on the red diamond once again as requested.

"Thirty-one, passe, impair, noir." This time, Gombaud has lost, and the croupier rakes in his chip, along with the chips of all the other players who have made losing bets.

"200 livres on red." Gombaud proceeds as Pascal had told him to: "Keep on doubling up after you lose, until you win. That is the strategy. And when you win, then you should go home."

"Seventeen, manque, impair, noir."

Once again, Gombaud has lost, and now he has to double his stake again. "400 livres on red."

"Certainly, monsieur," answers the croupier and then, when the ball has dropped into a slot on the wheel, he calls out calmly: "Six, manque, pair, noir," and gathers in Gombaud's 400 livres.

Only when he finally wins can Gombaud stop doubling his stake: "800 livres on red."

"Twenty-six, passe, pair, noir"—again, he has lost.

"1600 livres on red."

"Thirteen, manque, impair, noir"—his run of bad luck is simply never-ending.

Gombaud has no more chips left. He rummages in his wallet and pulls out a few Louis d'or, gold coins depicting the king, worth 3,200 livres. "All of this on red," he says, and the croupier swiftly changes the splendid coins into cheap-looking chips, which he then places on the red diamond, while the ball lands in the slot marked thirty-five. And thirty-five is black.

The other players at the roulette table observe the scene with growing interest. They stare in amazement as Gombaud, outwardly casual but in reality nettled by the situation, pulls the enormous sum of 6,400 livres from his wallet and bets it all on red. Only the croupier remains as calm as ever. "No more bets," he calls, as the ball starts to slow down and then finally falls into a slot on the roulette wheel. "Three, manque, impair, rouge." At long last, red has come, and the croupier uses his rake to push chips worth 6,400 livres toward Gombaud, who now has a total of 12,800 livres, including his stake. It is a colossal amount of money. He hurries to the cashier to have his chips changed into money, calculating as he goes. How much money had he bet in total at the roulette table from the moment he started again from the beginning? First it was 100 livres, then 200, then 400, then 800, then 1,600, then 3,200, and finally 6,400, making a total of 12,700. Although envious eyes followed him when he left the table with the 12,800 livres, his net profit turns out to be a mere trifle in his eyes—100 livres, the same amount as the stake with which he began.

"With my system, you can't expect to win much," Pascal cautions, when Gombaud comes to Port Royal des Champs a few days later to visit him and get the disappointment at his meagre winnings off his chest. Despite his now regular visits, Gombaud has still not gotten used to the convent's stale air, its bare walls, and the pallid nuns who live there. His desire to understand the game of chance, however, forces him to overcome his aversion.

"The fact that the net profit is so low—always the same as the stake with which one begins—is not the only thing that bothers me,"

Gombaud complains. "Playing like this is also dreadfully boring. Everything about games that inspires me, the appearance of the unexpected, a brilliant idea how to use some trick to snatch victory, even though you have almost given up hope yourself—all of that is missing. I admit that the amounts staked quickly increase when you double them, but that makes no difference to me with my wealth. Whether it's a hundred livres or a hundred thousand—if I know for certain that I will definitely win in the end, the risk that I love so much is absent."

"And then there is all the time that you waste at the gambling table for a sum that is, in your eyes, insignificant," agrees Pascal, who himself hasn't sacrificed even a moment for a game of chance since entering the solitude of the convent.

"What if, instead of always betting on red, I always bet on zero?" asks Gombaud, referring to the slot on the roulette wheel marked zero. As he says this, he begins to think that he has discovered a strategy that is better than that of the clever but unworldly Pascal. "Then, if I should win, I would get thirty-five times my stake. The game would still be just as tedious as before, but at least my winnings would be significant."

"But just think how low the probability is that zero comes," counters Pascal. "It's one in thirty-seven, which is just over 2.7 percent. It can easily occur that the ball doesn't land in zero for forty spins in a row. You mustn't discount that possibility."

"Considering how much money I have, that doesn't seem to be a problem."

"Monsieur Gombaud," protests Pascal, "even if you start off with one livre and double up forty times, you get more than a trillion livres—a million times a million livres!"

"That is indeed more than I have," admits Gombaud sheepishly.

"And certainly more than the casino has, also," adds Pascal, and Gombaud swiftly confirms this:

"Certainly more than those poor devils. Because their finances are on an insecure footing, the casinos have introduced a limit. For example, I can't bet more than a hundred thousand livres on red. It seems that the casinos can't afford to pay out more than a hundred thousand livres in winnings to any one individual."

"I wasn't aware of that," replies Pascal, and after a brief moment of thought, he continues: "It's not because they are poor that the casinos set a limit—it's because they are clever. By doing so, they destroy my strategy of doubling up after losing. For, Monsieur Gombaud, if you begin with a stake of a hundred livres, keep betting on red and, as luck would have it, black wins eleven times in a row, then you are forced to stop playing without winning. If you keep doubling up from 100 onward, you get 200, then 400; 800; 1,600; 3,200; 6,400; 12,800; 25,600; 51,200; and 102,400—i.e., more than a hundred thousand livres after the tenth doubling up, and a sum that you are not allowed to bet."

"Black eleven times in a row—when could that ever happen?" protests Gombaud. Even today, many people who are in thrall to the roulette table hold the mistaken belief that there is a "law of balance." Intently, they study the display boards that the casinos have set up by the tables, seemingly as a service to the customers, showing the winning numbers of the previous spins. When those who believe in the "law of balance" notice that a black number has come five times in a row, they will hurry to bet large sums on red, believing that the balance between red and black must be reestablished.

Rather than giving a theoretical explanation, we can illustrate much more clearly how dangerously wrong this assumption is in the short and medium term with a true story, as told in Rolf Dobelli's bestseller *The Art of Thinking Clearly*:

> In the summer of 1913, something incredible happened in Monte Carlo. Crowds gathered around a roulette table and could not believe their eyes. The ball had landed on black twenty times in a row. Many players took advantage of the opportunity and immediately put their money on red. But the ball continued to come to rest on black. Even more people flocked to the table to bet on red. It had to change eventually! But it was black yet again—and again and

> again. It was not until the twenty-seventh spin that the ball eventually landed on red. By that time, the players had bet millions on the table. In a few spins of the wheel, they were bankrupt.[1]

"Forget my system," Pascal beseeches the dismayed Gombaud. "Forget all systems for winning at roulette. If you are addicted to gambling at the casino, then follow this one iron rule alone: only play with the chips that you receive for your money when you enter. View this money from the outset as money down the drain, the loss of which does not bother you. And do not exchange any more money for chips during the course of the evening. Enjoy playing—you obviously can, although I would not. And, while you are playing, do not think about whether you will be going home at dawn having won or lost. It would only spoil your enjoyment."

The system for winning devised by Pascal must have given the Swiss mathematician Nikolaus Bernoulli the idea for an unusual game, almost exactly fifty years after Pascal's premature death. Nikolaus Bernoulli was the nephew of Jakob Bernoulli, the author of *Ars Conjectandi,* about which we wrote in chapter 4, where we mentioned that Nikolaus Bernoulli had published the book after the death of his uncle. In a letter to his colleague Pierre Rémond de Montmort, who was also an enthusiast for probability and games of chance, Nikolaus Bernoulli wrote about the invention of his curious game. The letter would have been forgotten if a cousin of his, Daniel Bernoulli, who was employed by the tsar as a professor of physics in St. Petersburg, had not rediscovered the game described in the letter and then presented it to the public in his own words. Daniel Bernoulli takes us into a St. Petersburg casino, where an enterprising croupier—we'll call him Alexei—is proposing the following idea to the director of the casino:

"We could demand a large sum of money from the players, let's say a hundred rubles, if he wants to play the following game with roulette. Once he has paid this fee, he can then decide if he wants to bet on 'permanence red' or 'permanence black.' Let's assume he bets on permanence red. If the ball lands on black, the player receives one ruble, and the game is over. If the ball lands on red, the wheel is spun once more. If the ball lands on black the second time, the player receives two rubles, and the game is over. If the ball lands on red, the wheel is spun again. If the ball lands on black on this third spin, the player gets four rubles, and the game is over. The slot with the number zero is covered so that the ball cannot drop inside it. And so the game goes on—every time the ball lands on red, the wheel is spun again. As soon as it lands on black, the player receives a payout and the game is over. The game's attraction for the player is that the payout is doubled each time, beginning with one ruble for the first spin, two rubles for the second, then four, then eight, then sixteen, then thirty-two, then sixty-four, and so on. If the ball lands ten times in a row on red, the player can expect to win 512 rubles, more than five times his initial stake."

"Good for the player, bad for us," replies the director, much to the surprise of the inventive croupier, who is now no longer completely sure of his idea. "What made you come up with the amount of a hundred rubles as a fee?" the director asks Alexei.

"To be honest, I thought that would be an amount that would encourage *vabanque* players to get involved."

"And we would certainly lose in the long term," answers the director with a scowl. Rapid mental calculations are a passion of his, and are also helpful in his line of business, but for Alexei's sake, he explains his dislike of the proposed game as comprehensively as he can: "Let's assume that about a million players, to be precise 1,024,000 players, are interested in your wonderful game. And let us assume that we ask for five rubles, not a hundred, as a fee."

"Why 1,024,000 players?" Alexei wants to know.

"Because the number 1,024 can be divided by two ten times in a row without any remainder. You'll soon see what I mean. In any case, each of these players pays the deposit. The casino therefore has more than five million in the bank—5,120,000, to be precise. To keep things clear and simple, let us assume that all of the players bet on 'permanence red.' And let us imagine that we run this game for each player individually. I will now create a table and write in the first column what color the ball can land on in sequence; in the second column how many of the million-plus players to whom the color in question applies, according to the law of probability; in the third column how many rubles each of the players thereby receives; and in the fourth column how much the casino has to pay from its funds. You will soon understand what I mean."

"I suspect that the fourth column will contain the product of the two numbers in the second and third columns," says Alexei, showing that he is a quick learner.

"Exactly. On the first line, I'll write on the left the letter B for the event that the ball lands on black the very first time. If the slot with zero is blocked for the ball, the probability of black is one-half. In such a case, we have to pay the player one ruble. That means that we have to pay just over half a million of the players"—the director writes "512,000" in the second column—"one ruble."—He writes "1" in the third column. He writes "512,000" in the fourth column and says, "These players are now out of the game and there is still about 4.5 million rubles in the bank."

Then he continues: "On the second line, I'll write on the left the two letters RB for the event that the ball lands first on red and then on black. The probability of this is half of one-half, so a quarter. In this case, we have to pay the player two rubles. That means that we have to pay just over a quarter of a million of the remaining players"—the director writes "256,000" in the second column—"two rubles."—He writes "2" in the third column. In the fourth column, he again writes "512,000" and explains: "256,000 times 2 is 512,000. There is now about four million rubles left in the bank. On the third line, I'll write on the left the three letters RRB for the event that the ball first lands twice on red and then on black. The probability of this is half of one-quarter, that is, an eighth. In this case, we have to pay the player four rubles. That means that we have to pay just over an eighth of a million of the remaining players"—the director writes

"128,000" in the second column—"four rubles."—He writes "4" in the third column. He again writes "512,000" in the fourth column and explains: "128,000 times 4 is again 512,000. Now there is just over 3.5 million rubles left in the bank. And so it relentlessly continues."

Without further ado, the director extends his table by seven further lines. In the first column is written, one under the other, B, RB, RRB, RRRB and so on up to RRRRRRRRRB. Next to each of these are written the numbers 512,000; 256,000; 128,000; 64,000; 32,000; 16,000; 8,000; 4,000; 2,000; and 1,000. Next to each of these in the third column is written 1, 2, 4, 8, 16, 32, 64, 128, 256, 512; and in the fourth column is written ten times underneath one another the enormous sum of 512,000 rubles.

1,024,000 players, each paying a fee of 5 rubles
Total casino funds: 1,024,000 x 5 = 5,120,000 rubles

How the ball lands:	Applies to:	Rubles per player:	Total rubles:
S	512,000	1	512,000
RS	256,000	2	512,000
RRS	128,000	4	512,000
RRRS	64,000	8	512,000
RRRRS	32,000	16	512,000
RRRRRS	16,000	32	512,000
RRRRRRS	8,000	64	512,000
RRRRRRRS	4,000	128	512,000
RRRRRRRRS	2,000	256	512,000
RRRRRRRRRS	1,000	512	512,000
...	...	...	???

Figure 6.1: The table that the casino director uses to explain the so-called "St. Petersburg paradox" to his croupier.

"Look at the table, Alexei! From the tenth line downward, we have none of the five and a bit million rubles in the bank. It has all already been paid out as winnings to the players. And yet there are still a thousand lucky players for whom chance has reserved even greater winnings than 512 rubles. Five hundred will get 1,024 rubles, 250 will get 2,048 rubles, 125 as much as 4,096, and so on. We won't be able to pay out these winnings, because we will long since be broke."

"But I started off with a fee of a hundred rubles, not five," protests Alexei.

"That makes no difference to the principle of my loss account. I would only need to start with many more players and include many more lines. In the end, the total of the amounts in the fourth column will exceed the funds that we have stashed away."

"Wouldn't a thousand rubles be enough as a fee?" Intimidated by the director's verbal onslaught, Alexei makes one final attempt to salvage his idea.

"Forget it! Even if we demanded a million rubles as a fee, we would be sure to lose out with this game in the long run. I've shown you the calculations. And anyway—I can't imagine that there is anybody foolhardy enough to get involved in this game for the massive sum of a thousand rubles. The risk would seem too great."

Full of disappointment, Alexei leaves the director's office. He is not only disappointed but also confused, for he senses a paradox here: on the one hand, the director makes it quite clear that only a madman would sign up for the game with a fee of a thousand rubles. If we suppose that he bets on "permanence red" and, in the first spin, the ball lands on black, which would happen on average in half of all cases—in such a scenario, the player suffers the disastrous loss of 999 rubles after just a few seconds. Viewed like this, it is even doubtful that one could tempt many speculators to play the game with a fee of only a hundred rubles.

On the other hand, the director has rejected the game because he has worked out that the casino would certainly lose with this game in the long term, and this regardless of how much each player's fee is. How can it be that nobody would be prepared to risk playing a game where the casino is bound to end up on the losing side for a change?

Because Daniel Bernoulli sets this scene in a casino in St. Petersburg, people talk of the "St. Petersburg paradox" when referring to this inconsistency that has thrown Alexei into such confusion.

Were Pascal still alive and were Alexei, in his disappointment and confusion, to put the St. Petersburg paradox to him, he would most probably resolve it as follows:

"The way you have invented the game, my dear Alexei, it ends when, if the player has bet on 'permanence red,' the ball lands on black for the first time. That can happen in the very first round—which would be bad luck for the player—but it can also take many, many rounds. There is no fixed number of rounds by which one knows for certain that the ball must land on black. But nobody would stand for ever at the roulette table—at some point, one gets tired and leaves the casino. At some point, the wheel stops gliding evenly in the bowl or the ivory ball breaks up. The idea of an infinitely long game is an illusion, so one will have to set a limit. Consequently, one should adjust your game as follows: the casino sets a limit in advance, let's say a thousand rubles. The game proceeds as you have described, to begin with: after paying the fee, the player decides whether to bet on 'permanence red' or 'permanence black.' Let us assume that he bets on 'permanence red.' As the croupier, you spin the ball in the bowl until it first falls on black or—and this is the point—until the casino is forced to pay out the amount set as the limit. In this way, the game is guaranteed to end at the latest when this limit is reached. In all other respects, it remains as you have suggested: if the ball lands on black in the first round, the player receives one ruble and the game is over. If the ball lands on black in the second round, the player receives two rubles and the game is over. If the balls only lands on black in the third round, the player receives four rubles and the game is over. With each further round, the amount of money to be paid to the player is doubled if the ball lands on black for the first time in that round. But once this amount exceeds the set limit, the game is in any case over—the player receives the amount of money set as the limit and has fully won the 'St. Petersburg roulette,' as we might call it."

"But nobody would want to play the game for a fee of a thousand rubles if the limit is also only a thousand rubles," Alexei objects.

"You are absolutely right," answers Pascal, and continues his explanation: "Let us put ourselves in your director's position and consider how much, over the course of numerous games, the casino would pay out on average to each player. There's one ruble for half of all players, the unlucky ones for whom the ball lands on black in the very first round. Then two rubles for a quarter of all players, for whom the ball lands on red in the first round, but on black in the second. Then four rubles for an eighth of all players, and so on and so forth, until the limit has been reached. Before the thousand-ruble limit has to be paid out, it is first necessary to pay out 1, 2, 4, 8, 16, 32, 64, 128, 256, and 512 rubles respectively to half, a quarter, an eighth, and so on, of all players. And a thousand rubles—or, if we are generous, 1,024 rubles, since 1,024 is twice 512—has to be paid out to those players remaining at the end, those few players—on average, only one in 1,024—who win the full St. Petersburg roulette. As the director correctly calculated, this payout is the same as if he were to take half a ruble times the number of players from the bank's funds as often as there are winning possibilities before the limit is reached. That is—if I count off how many numbers 1, 2, 4, 8, 16, 32, 64, 128, 256, 512 there are—exactly ten winning possibilities. In addition, there is a whole ruble for the few lucky players who win the full St. Petersburg roulette. Since ten times half a ruble is five rubles, and plus one ruble is six rubles, it would only be fair if the casino demanded a fee of six rubles from each player for the St. Petersburg roulette, assuming a limit of a thousand rubles."

"At six rubles, I can well imagine that lots of players might be tempted to take part in this game," admits Alexei, but then adds: "But the casino wants to win in the long term. Would it not be wiser to set the deposit at seven rubles?"

"It's certainly possible. But even with a deposit of six rubles, there is a very sophisticated trick to guarantee that the casino wins, my dear Alexei. You simply do not cover the slot with the number zero, unlike in your original suggestion, so that the ball can also fall into this slot. If zero comes, it is agreed that all subsequent payments to the players are halved. The casino uses a similar trick to win with normal roulette."

"And what if one raises the limit, let's say to a million rubles?"

"The increase to the fee is negligible. For a limit of a million rubles, the fee is just eleven rubles. You can work it out yourself."

"Then I don't understand why the idea of a game without limits is so paradoxical."

"The director is right. It's pointless. Forget it!"

A whole series of mathematicians, beginning with Daniel Bernoulli, refused to take Pascal's last piece of advice seriously. They used all sorts of tricks in an attempt to reveal the secret of the paradox in St. Petersburg roulette without a limit, but with questionable success. One of those who tried their hand at the St. Petersburg paradox was Karl Menger. He believed he had got to grips with it with the help of his father's marginal utility theory—remember that, for a billionaire, winning a million is not worth mentioning, whereas for a pauper, winning a million is a life-changing event. In the course of his deliberations, he recognized that the paradox can only be avoided by closing the door on the possibility of infinite winnings. As a student, he had already written down his reflections on the topic in an essay, but it was only in 1934 that he managed to get this paper published in the *Zeitschrift für Nationalökonomie.*

That same year also saw the publication of a book by Karl Menger that laid the foundations for the study of the nature of games other than mere games of chance. Therefore, in the next chapter, we will return once more to him and his teacher Hans Hahn.

7
PLAYING WITH SCHOLARS

VIENNA, BETWEEN 1921 AND 1934

"This book is a disappointment."

Hardly has he spoken these words and noticed the dismay of his colleague and former pupil, than Hans Hahn wishes he could take back the harsh expression "disappointment." But the word has already escaped through the "barrier of his teeth," as the ancient Greeks would say. An icy silence pervades the room, and Karl Menger says nothing in reply.

"Menger, don't get me wrong," says Hahn placatingly. "The book is too mathematical for a sociologist, economist, or philosopher, while at the same time too vaguely written for a mathematician. The title, *Morality, Decision and Social Organization,* promises a great deal, but I fear that most readers will not find in the book what they are expecting from the title."

His words are followed by a tense silence.

"By the way, Menger, how are you?" Hahn realizes that it is pointless to speak further about Menger's book, and so he switches the topic to his young colleague's health, although his enquiry is more than just a social courtesy.

Karl Menger had been taken seriously ill shortly after attending the first few of Hahn's seminars on the nature of a curve. In that year of 1921, a time of hardship due to the deep scars that the Great

War had left on the country, it was scarcely surprising that he should be struck down by tuberculosis, known as the "Viennese disease" and widespread at the time. Menger was forced to spend months on a cure, taking refuge in Aflenz, a village nestled high up in the mountains near the pilgrimage site of Mariazell, with its fresh and fragrant air. Here, he continued his mathematical studies, using only the books and writings that he had brought from Vienna or which were sent to him. During his enforced residence in Aflenz, his father passed away in Vienna. Old Menger had intended to republish his 1871 work *Principles of Economics* in a comprehensively reworked second edition, but his death ended these plans. Young Menger had the manuscript sent to him and, in memory of his deceased father, set about completing the work old Menger had begun. When he finally had the finished book in his hands in 1922, he had become an expert on economic laws.

Letters between Aflenz and Vienna maintained young Menger's contact with the Viennese scene, and he kept up his regular correspondence with his Viennese friends and colleagues even when, after almost two years of convalescence in Aflenz and after completing his mathematical studies in Vienna, he went to work under Brouwer in Amsterdam. Above all, he corresponded with Hilda Axamit, a student of actuarial mathematics, whom he called "Mitzi" and who would later become his wife in 1935.

There was indeed a great deal to write about in the Viennese academic world. The tireless Hans Hahn had gathered around himself some of the cleverest people at the University of Vienna who shared his interests in physics and philosophy, interests that had been aroused just after the turn of the century, when Hahn himself was studying at the university. The two most influential scientists whom he had gotten to know at that time were Ludwig Boltzmann, the inventor of statistical physics, and Ernst Mach, whose fame as a physicist derived above all from his ability to describe with great precision the ballistic shockwaves caused by bullets traveling faster than the speed of sound. Mach was also influential because he attempted with uncompromising rigor to reduce physics to what could be directly observed. He considered discussions about the truthfulness of theories to be superfluous, since he didn't think that "truth" existed, as such. Usefulness alone was what mattered, as well as the congruence with experiments. He had no time for Ludwig Boltzmann's theories, since these were based upon the assumption that there were such things as atoms and groups of atoms, so-called molecules. Mach, however, considered atoms to be figments of the imagination, dreamt up by dodgy chemists.

Boltzmann would go on about Mach's narrowmindedness with biting irony, but, if anything, this only encouraged Mach in his opinions. When one of his students was being tested and mentioned atoms, Mach would fix him with a remorseless gaze and hiss in a thick Viennese accent: "Have you ever seen an atom?"

"No, Professor."

"Then get out of here and start looking for one! And only come back when you have found one and can show it to me."

The irony of this anecdote is that Mach and Boltzmann, despite the strict dividing line between their ideas, valued each other greatly as people.

Both the merciless rigor of Mach and the imaginativeness of Boltzmann impressed Hans Hahn, but he soon found himself without a philosophical mentor in Vienna: Mach, who had suffered a stroke some years previously, left the city shortly before the outbreak of World War I, while a few years before his departure, on September 5, 1906, Boltzmann, plagued by the deepest depression, hanged himself from the crossbar of a window casement in Duino. Although Boltzmann and Mach almost never shared the same opinion, Hahn managed to draw a single conclusion from their different ways of thinking: the need to go beyond the ideas of the philosophers venerated in Vienna at the time, such as Kant, Schelling, and Hegel. No matter how important it is to uphold the traditions of ancient philosophy, scholasticism, and patristics, it is equally important to introduce new and modern schools of thought. It is the schools of thought that we owe to Boltzmann and Mach that result from the fascinating discoveries of mathematics and the exact sciences.

Until his death, Hahn considered it a matter of paramount importance to overcome speculative idealism. In his view, only theorems that could be experimentally tested and experienced, paired with logic and mathematics, could form a sound, philosophical foundation. He felt moderately in tune with the English empiricists John Locke and David Hume. He admired the incisive Bertrand Russell, however, whose *Principia Mathematica*, coauthored with Alfred North Whitehead, had just been published and aimed to reduce the whole of mathematics to a limited set of basic rules, praising him as the "most important philosopher of our time"—an unconventional judgment in Vienna at the time, with Russell practically unknown at the Institute of Philosophy.

During the final years of peace before war broke out in 1914, Hahn would meet up with kindred spirits in the coffeehouses that were then to be found on every corner of the magnificent Ringstrasse boulevard, the gathering constituting an informal philosophical circle. Richard von Mises, the physicist, mathematician, and brother of the theoretical economist Ludwig von Mises, was one member of the circle, and Philipp Frank, a brilliant young physicist, completed the philosophizing trio. Otto Neurath, a friend of Hahn's and later his brother-in-law, would join them from time to time. The men saw themselves as the three musketeers of a new and exact philosophy, with Hans Hahn as d'Artagnan. But before a serious philosophical school could develop out of their coffeehouse chats, the long war, with all its devastation and upheaval, got in the way.

Hahn didn't give up, however. After the war, and after he had taken on the mathematics chair at the University of Vienna, the chair in philosophy, which had once been held by Mach and then Boltzmann after him, became vacant. Hahn pulled out all the stops at the Ministry of Education in order to tempt what he considered to be the best man to come to the capital of Austria, now reduced to a small country. Moritz Schlick seemed to be the most suitable candidate for the job. He was on good terms with Einstein, while Max Planck, the founder of quantum theory, was his doctoral supervisor. And indeed, the tall, athletic Hahn, together with his powerfully built brother-in-law Neurath, an economist whose ideas were very close to those of socialism, and the newly appointed professor of philosophy Schlick, always elegantly dressed and blessed with natural dignity, did succeed in forming a new trio that accepted into their circle young scientists with an interest in modern philosophy. "Circle" is the right word here, for the intellectuals who met each Thursday evening in the gray building at Währinger Strasse 38–42, in a small lecture hall that was normally reserved for meteorology, called themselves the "Vienna Circle." This is not the place to reel off a complete list of those accepted into the Vienna Circle. Suffice it to say that Hans Hahn ensured that his two most gifted former students, Kurt Gödel and Karl Menger, were regular attendees at these Thursday meetings.

Hahn believed that the Vienna Circle's task was to reestablish philosophy based on empirical observation in conjunction with logic and mathematics, and his role model was still Bertrand Russell. But, one Thursday evening, Kurt Reidemeister, a professor of geometry and a member of the Vienna Circle, shows a slim volume to those present: "Gentlemen, I believe that we should concentrate our discussion on this book: the *Tractatus logico-philosophicus* by Ludwig Wittgenstein."

"Ludwig who?" asks Neurath.

"Wittgenstein," repeats Reidemeister. "A name of some renown. His father Karl Wittgenstein was the Austrian equivalent of Krupp—a steel magnate of boundless wealth." Otto Neurath pulls a disapproving face, but Reidemeister carries on, still facing him: "He had many children. His son Ludwig wanted none of his inheritance, gave it all away to his relatives, and now lives like a monk. He even refused to let his family provide the money necessary for printing his book." Neurath still seems unimpressed by this, so Reidemeister now turns to Hahn: "Do you know who has sung the book's praises and edited a dual-language version, in English and German? Bertrand Russell."

"Indeed?" Hahn is duly impressed, but Neurath flicks through the book and stops at the last few pages, before saying, "Let me read this out: 'The feeling of the world as a limited whole is the mystical feeling.'[1] Gentlemen, I am almost certain that this book is about metaphysics."

When something is called "metaphysics," this equates to a death sentence in the Vienna Circle. For it is the unspoken consensus of all the members that "behind" physics—and that is the literal meaning of "meta"-physics—there is absolutely nothing.

Schlick takes the book out of Neurath's hands and, after a quick look, comments: "Well, I can see that this Wittgenstein fellow can at least structure his thoughts clearly: every statement is numbered. The most important propositions have just one figure as a number, while the sublevels which elaborate on these important principles are marked with subnumbers. This seems to be used systematically and will certainly be of great assistance to us when discussing the book. I too would like to read out a sentence, one from the preface, where Mr. Wittgenstein says: '[The book's] whole meaning could be summed up somewhat as follows: What can be said at all can be said clearly; and whereof one cannot speak thereof one must be silent.'"[2] Schlick now addresses Neurath directly: "That is almost exactly the motto of our group." His voice becomes noticeably louder in order to lend weight to his words. "It is my opinion, if the book itself lives up to this sentence in the foreword, and if Russell praises it to the skies and rates it as being of considerable importance, that it is certainly worth our effort analyzing it—naturally, my dear Neurath, with the utmost rigor."

Hahn is only too happy for Schlick, with his natural authority, to make this decision. He could not have put it any better himself. And he has a premonition of what would indeed turn out to be the case: for years afterward, Wittgenstein's *Tractatus logico-philosophicus* was to become the mainstay of the Vienna Circle. Despite this, however, the myriad attempts to get the author himself to join the Vienna Circle all failed. "Wittgenstein declared to me that everything that there is to say in philosophy is said in his book," reports Friedrich Waismann to his colleagues in the Vienna Circle, and goes on, full of disappointment, "He considers our endeavors to be a waste of time, speaks disparagingly of 'Hahn and Neurath and all that clique' and refuses to come under any circumstances. He showed me the door with abrupt curtness."

Karl Menger's relationship with the Vienna Circle was completely different from Kurt Gödel's. Gödel was a regular participant at the meetings, and all those present could sense his unwavering interest in the topics discussed. Yet he himself never spoke. His silence had a hidden explanation: he would privately attempt to distill from all contributions to the discussions that which he could translate into his purely mathematical terms. Sometimes, that was much less than what was contained in the Vienna Circle's sometimes heated debates about Wittgenstein's book—Gödel's respect for Wittgenstein was minimal. Sometimes, though, Gödel did reflect on much more than what was allowed in the Vienna Circle, matters that would have been damned as "metaphysics." It was only decades later that Gödel lifted the veil on his thoughts somewhat: in 1941, newly arrived as a refugee from Hitler in Princeton, he translated into the language of mathematics the most metaphysical of all "proofs of the existence of God," the so-called "ontological argument" by the medieval monk and thinker Anselm of Canterbury. In doing so, he knew full well that it would have no effect on any person's religiousness or irreligiousness. Sometimes, Gödel claimed that his argument for the existence of God was merely an insignificant mathematical game for him, a means of putting his talent to the test. But perhaps that was just a smokescreen that he used to conceal his religiousness. Had Neurath or other unrelenting advocates of the ban on metaphysics guessed the nature of Gödel's ideas, he would have been immediately excluded from the Vienna Circle.

Karl Menger, meanwhile, supported as best he could the Vienna Circle's endeavors, particularly those of Hahn and Schlick. He was not as regular an attendee of the meetings as Gödel, because he was doing research at Harvard University in Massachusetts and at the Rice Institute in Texas in 1930 and 1931. He did, however, keep himself up to date via letters from a pupil of his in Vienna, Georg Nöbeling. With great enthusiasm and commitment, he backed Hahn's suggestion of bringing the themes discussed by the exalted members of the Vienna Circle to the wider public. The aim was to organize five or six public lectures each year in the large mathematics lecture hall, to be given by eminent authorities and open to all those who were prepared to pay the equivalent of the ticket price for the State Opera House, Vienna's cultural temple. Under the general heading of "Old Problems—New Solutions in the Exact Sciences," the speakers included notable figures such as the pioneer in polymer chemistry Hermann Mark, the young winner of the Nobel Prize in Physics Werner Heisenberg, Vienna's most eminent theoretical physicist Hans Thirring, and also renowned members of the Vienna Circle itself, above all Hans Hahn. The enormous lecture hall was always packed to the rafters for these lectures, despite the entrance fee and the fact that, in those lean years of economic crisis and mass unemployment, the price of an opera ticket was no trifle.

Menger, well versed in matters of economics, lent his support to this idea and was in charge of the takings, the aim being to erect a statue on Boltzmann's grave and also to keep the mathematician Olga Taussky, who carried out administrative tasks for Hahn and Furtwängler, financially afloat, since paid posts at the university were rare. Menger was experienced in the organization of large-scale lecture events because he was in charge of a "Mathematical Colloquium," which presented the latest findings of mathematics to a specialist audience. Despite his strained relationship with Brouwer, Menger also invited his former mentor to give a talk in Vienna.

"This is our chance to tempt Wittgenstein out of his reclusion," suggests Schlick enthusiastically. "Waismann and I once managed to talk to him briefly, when his sister arranged a meeting. While he still doesn't want to hear anything about philosophy, if we tell him that Brouwer is going to talk in Vienna about the foundations of mathematics and not about philosophy, I believe that there is a possibility of his coming."

A delegation of members of the Vienna Circle go to see Wittgenstein, who fortunately does not dismiss them out of hand, and they report back to the others that Wittgenstein has agreed to give careful thought to whether he will come or not.

In the moments before Brouwer's lecture, Hahn and Menger wait nervously outside the hall as it gradually fills up. Suddenly, they see Wittgenstein walking slowly down the long corridor, very much in the manner of the wise and austere Tolstoy. Delighted, Hahn welcomes him with open arms, before leading him into the hall and offering him a place of honor in the first row—which Wittgenstein vehemently rejects, seating himself modestly in the fifth row instead. Kurt Gödel is also there, perched at the other end of the hall. Each of the two men, Wittgenstein and Gödel, is fascinated in his own way by Brouwer's lecture. Gödel draws conclusions from it that he can link up with his formal thinking in such a way as to make him the greatest logician since Aristotle. Wittgenstein, on the other hand, is so impressed that he even attends the inevitable gathering afterward in a nearby coffeehouse and, for the first time in years, speaks once more about philosophy—something he will not now desist from doing in Cambridge until the end of his life.

Brouwer's lecture was thus one of the major successes that Menger and Hahn achieved for the Vienna Circle. From then on, the members cherished the illusion that Vienna would regain its pre-war glory as a city of science, and that it would rise to become the center of the new sciences. But the political events of the years that followed, culminating in Hitler's rise to power in Germany and the unmistakable enthusiasm about this among large sections of the Austrian population, unsettled the extremely sensitive Menger. He increasingly withdrew from the university, where many students and professors had been taken in by the National Socialist propaganda, as well as from the Vienna Circle itself. The pamphlet *Wissenschaftliche Weltauffassung* (Scientific World-Conception), published by several members of the Circle, was not to his liking. When people asked him if he was a member of the Vienna Circle, he would reply: "To be honest, I'm not a member of the Vienna Circle. I see myself as being merely related to it."

Menger sought to dispel his gloomy mood, a trait he had inherited from his father, in the summer resorts around Vienna or in the picturesque landscapes of the Salzkammergut region. He failed. He sensed that the pervasive political trends were at odds with the ideals to which he himself felt committed. And he regretted that the members of the Vienna Circle, who were increasingly shutting themselves off from the outside world, did not recognize the things that it was essential to discuss in this time of political upheaval.

Menger is convinced that the actions and agitations of the political factions must be explored using the method of precise thinking. What motives drive people to say things or to take action? What goals do they have in mind? What norms are they based on? How can one pose ethical and moral questions in such a precise way that a clear approach is laid out for the answers? He draws sketches that illustrate his deliberations: circles represent the individual social groups that are held together by common interests; arrows between these circles, drawn with varying degrees of thickness and pointing in various directions, show to what degree the individual groups collaborate or fight against one another. He gives the title *Moral, Wille und Weltgestaltung* (Morality, Decision and Social Organization) to the book he writes in that year of 1934—and heavy criticism, above all from the members of the Vienna Circle, is not long in coming.

"Menger, I fear that the thin air on Semmering"—a mountain one hundred kilometers south of Vienna that was then a popular summer resort for the Viennese—"didn't do you much good when you were writing this book."

"Menger, aren't you aware that Bertrand Russell doesn't believe that morality and ethics have anything to do with a theory of knowledge?"

"Menger, I can detect in this book a major dose of metaphysics."

And finally the harsh criticism voiced by his former teacher, Hahn: "This book is a disappointment."

One man, however—not a mathematician, but rather an economist, a pupil of Ludwig von Mises and detached from the Vienna Circle, and the successor to Friedrich von Hayek as the Director of the Austrian Institute of Economic Research—does recognize the groundbreaking nature of Menger's book. For him, the book is a revelation. His name is Oskar Morgenstern.

8
PLAYING WITH TWO CARDS

PRINCETON, NEW JERSEY, 1938

"A game of poker with only two cards? That's ridiculous!"

Perplexed, Oskar Morgenstern stares at John von Neumann, who calmly carries on talking: "Please be patient. You will soon see that I can thereby solve your conundrum."

The conundrum in question has been bothering Oskar Morgenstern for a long time. A keen reader of Arthur Conan Doyle, he cannot get a particular scene out of his mind, one where Sherlock Holmes, the ingenious detective, and Professor Moriarty, his archenemy, are traveling from London toward Dover.

Morgenstern recounts the story to von Neumann slightly differently from how Conan Doyle wrote it in *The Final Problem*—his version is even trickier than that of the famous crime author. "Holmes wants to escape from the armed Moriarty and so he gets on the Dover train at Victoria Station. When he looks out of the compartment window, he sees to his horror that Moriarty, just as the train is starting to move, manages to jump up and climb into a compartment. It isn't Holmes's compartment, and so he is safe from Moriarty during the journey, since the train is not an express, with a corridor linking all the compartments, and so the carriages are separated from each other. But when Holmes and Moriarty alight in Dover, the game will be up for Holmes: the villain will shoot him dead in cold blood on the platform.

"Nonetheless, there is a glimmer of hope. On its way from London to Dover, the train stops in Canterbury. If Holmes gets off in Canterbury and Moriarty happens not to be looking out of a window from which he could see Holmes alight, then Moriarty will continue on to Dover and Holmes will be saved. The only problem would be if Holmes gets off on the wrong side and Moriarty therefore sees him leave the train in Canterbury.

"But there is a second problem on top of that: Moriarty knows that Holmes has seen him get on the train. He will, therefore, wager that Holmes will leave the train early. Consequently, Holmes says to himself, Moriarty will not even bother looking out of the window in Canterbury and will instead get off the train and, as soon as it has departed again, will shoot Holmes. Holmes therefore deduces that it is wiser not to alight in Canterbury and instead to continue on to Dover, thus tricking Moriarty into staying in Canterbury.

"Holmes also knows, however, that Moriarty can follow these deliberations of his and will therefore come to the conclusion that it is after all more advantageous for him, Moriarty, merely to look out of the window in Canterbury and, if he doesn't catch sight of his enemy Holmes, to travel all the way to Dover. Thus, Holmes thinks to himself, getting off the train in Canterbury is the better option after all. But he knows that Moriarty knows that he is thinking this.

"Which of the two options, getting off the train in Canterbury or traveling on to Dover, is the wiser?"

While still in Vienna, Oskar Morgenstern had once taken up the book *Morality, Decision and Social Organization* by Karl Menger, in which Menger writes about, among other things, the mutual interaction of two groups that are hostile to each other. In this abstract framework, Morgenstern recognized the above dilemma facing Sherlock Holmes, and he saw that Menger was trying to provide in mathematics the tools to describe this and many other kinds of interaction. Although Morgenstern had not studied mathematics, he had always been an admirer of the subject. His ideal was to introduce mathematical rigor into economics, the specialist field that he taught at the

University of Vienna and in which he served as an economic advisor for Austria, which is why this book by the mathematician Menger was so important for him.

It was, however, no longer possible for him to visit Menger and talk about the book with him in person. Menger had already left Austria. The banning of the Social Democrats—a bitter blow for the Vienna Circle, the majority of whose members were close to the party—the bloody unrest in the country, and the seemingly inexorable rise of Hitler's supporters in Austria disquieted him. On top of all that, Hans Hahn had died in 1934, the year of the greatest turmoil, while undergoing an operation on his stomach, and the Ministry refused to appoint a successor to his chair. The final straw for the Vienna Circle, however, was the murder of Professor Schlick in the middle of the University of Vienna's main building two years later, when one of his former students, a mentally unstable man who may well have been secretly spurred on by Schlick's enemies and who had previously threatened his former teacher on several occasions, shot Moritz Schlick on the so-called Philosophers' Staircase.

"Leave Vienna at all costs," was Menger's thinking when he was offered the chair in mathematics at the University of Notre Dame in Indiana, although he didn't initially resign from his position in Vienna. It was only after March 1938, when Hitler and his troops marched into Austria, that he finally gave up his professorship in Vienna.

Morgenstern, on the other hand, remained in Vienna until 1938. His supporters Hans Mayer and Ferdinand Degenfeld-Schönburg were convinced that he and his expertise provided a valuable contribution to both the university and the state, and they shielded him against the hostile attacks of Othmar Spann, who plotted and schemed against Morgenstern. Born in the Saxon town of Görlitz, Morgenstern had been living in Vienna since his school days and felt a close bond with the city. It was a real stroke of luck for him that, just before Hitler marched into Austria, he happened to be on a lecture tour of the United States. He couldn't stand National Socialism, and so a return to Vienna was out of the question; besides, his name was on the Gestapo's blacklist. He therefore decided to remain in New Jersey, having been offered a position there at Princeton University.

It is almost seven hundred miles from Princeton University to the University of Notre Dame, where Menger is teaching. On the other hand, it is just one mile to the Institute for Advanced Study in Princeton. And that is where John von Neumann, the uncrowned king of mathematics, is tirelessly carrying out research. Back in Vienna, Eduard Čech, a mathematician from Brno, had told Morgenstern about John von Neumann, saying that he was sure to be open to puzzles of the kind that Morgenstern had in mind. So Morgenstern now goes to see von Neumann with Menger's book and his problem. What should Holmes do: get off in Canterbury or continue on to Dover?

"A fantastic conundrum!" says John von Neumann enthusiastically, after Oskar Morgenstern has told him his version of the Arthur Conan Doyle story. The two of them speak German with one another. It is, of course, Morgenstern's native language, and von Neumann can converse in half a dozen languages more easily than most people who have grown up speaking them.

"And it is a conundrum that doesn't only apply to crime stories," explains Morgenstern. "If I receive a bill, for example for a radio I have bought, I always wonder whether to pay it straightaway or to decide after trying it out whether to keep it or send it back. But waiting sometimes comes at a price, because some companies charge hefty reminder fees. That's another case where I'm not sure how best to proceed."

"Of course, of course, but let us put that example to one side for the moment and concentrate on poor Holmes's dilemma. After all, this problem occupies you the most. If I understand you correctly, you want to come up with a procedure that will show the participants in a given confrontation the optimal strategy for them. I had a look at this ten years ago. I wrote a six-page essay for the *Annals of Mathematics* on the theory of parlor games. I am sure that I have fully grasped the matter. You must simply view the whole thing as a game—whether it concerns Holmes and Moriarty, for whom it's a matter of life and death, or the mode of paying for or returning a product you have ordered, so that no financial loss ensues. In the game, there are two participants playing against each other. The game has rules that they have to follow. In the simplest case, they have one choice: should they make one move or the other? And they seek a strategy that they hope will help them win."

"I understand that, but what is the concrete procedure, for example in the case of Holmes and Moriarty?"

"See here, Morgenstern—Holmes and Moriarty are the two players, and they have to make their moves when the train stops in Canterbury. The possible moves are 'stay on' or 'get off.'" Von Neumann reaches for a sheet of paper and draws a table on it, writing "Holmes" on the left and then, to the right of this, "stay on" and "get off" between the lines. Then he writes "Moriarty" at the top and, beneath this, "stay on" and "get off" in the columns:

		Moriarty	
		stay on	get off
Holmes	stay on		
	get off		

"Now, we will write down what benefit the players gain from making the respective moves," continues von Neumann, and lists the individual cases: "If they both stay on the train, they will both arrive in Dover. That's bad for Holmes"—von Neumann writes "0" in the bottom left corner of the first box—"and good for Moriarty."—He writes "1" in the top right corner of the same box. "If Holmes stays on the train and Moriarty gets off in Canterbury, Holmes wins the game"—he writes "1" in the bottom left and "0" in the top right of the second box—"but Moriarty remains in Canterbury alone, like a fool. Now let's look at the scenario where Holmes gets off and Moriarty remains on the train and looks out of the window. In this case, in the bottom left box, I will write the figure ½ for both Holmes and Moriarty, since there is a probability of one-half that Holmes will be lucky and Moriarty will be looking in the other direction, but there is the same probability that Holmes will be unlucky and Moriarty will see him get off. Finally, both of them could get off in Canterbury, which would mean certain death for poor Holmes, so I'll enter the same numbers in the box as in the one that represents them both remaining on the train." Von Neumann now shows Morgenstern the completed table:

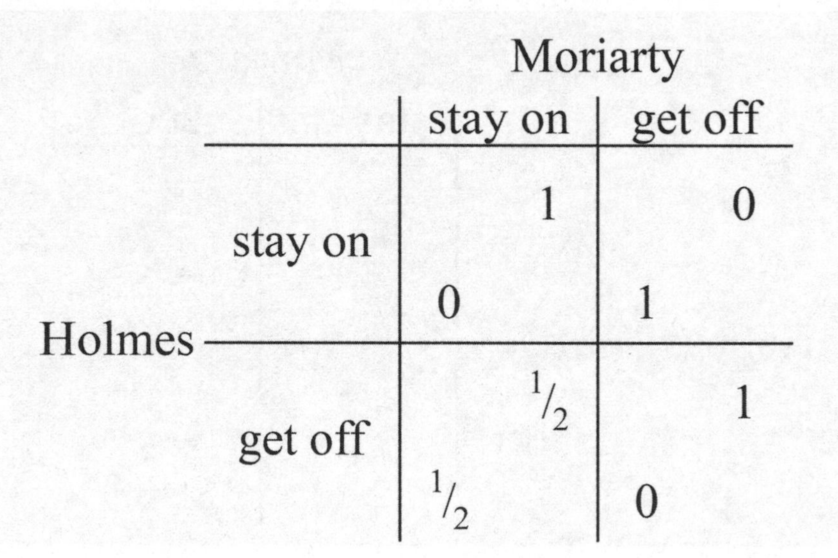

"Okay," says Morgenstern encouragingly, "But how do I get the solution to my conundrum from these numbers?"

"Did you know, Morgenstern, that I am a passionate poker player?" Morgenstern isn't quite sure why von Neumann suddenly poses this question. "No, why?" he asks, confused, but John von Neumann is already providing an explanation. "I usually lose at poker, but that doesn't matter to me. What counts is the allure of the game. It's too complicated for mathematical analysis with pen and paper, but if we concentrate on the fundamental aspect of the game, we can quickly see what its attraction is based on. Imagine that we are playing poker with just two cards: the ace of hearts and king of clubs."

"A game of poker with only two cards? That's ridiculous!"

Perplexed, Oskar Morgenstern stares at John von Neumann, who calmly carries on talking: "Please be patient. You will soon see that I can thereby solve your conundrum. It's obvious that the ace of hearts trumps the king of clubs. The size of the stakes is of no interest to us here. What matters to me is to find out the most advantageous strategy for each of us. Let's imagine that you shuffle the cards . . ."

". . . a rather onerous task with two cards," Oskar Morgenstern interjects.

"I admit that the whole thing sounds a bit crazy, but believe me it is not without appeal. Now I take a card. If I draw the king and am an honest chap, I declare the round to be over and I lose my stake. That's a point for you. If I draw the ace, I raise my stake and, if you match my stake, I win the game and get myself a point. And because the probability of drawing the king or the ace is one-half in each case, this means that, over a large number of rounds, if I play honestly and you match me when I increase my stake, each of us will, in the long term, get the same number of points. So far, so good. But the appeal of poker is that, after drawing a card, I don't have to be honest."

"If you hold the king in your hand, but still raise your stake, then you are bluffing."

"Correct. And now it's your turn. But you don't know which of the two cards I have drawn, because the other card is lying face down on the table. Now you can either match my stake increase or you can fold. If I do have the ace in my hand, that is, I am not bluffing, then folding is a better choice from your point of view. So you are due a point for that. If I am indeed bluffing and you decide to fold, then

I have won, despite the weak king in my hand, and this gives me a point.

"If you decide to match my stake when I am bluffing, however, you win the round—one point for you. Note that only if I raise the stake and you match me do I have to show the card I have drawn. And also consider the following point: I can decide even before the round whether—assuming I draw the king—to bluff or not. And you can decide before the round whether—assuming I don't concede defeat in the first place—to call or fold. And now"—John von Neumann reaches again for his pencil and paper—"I will show you what the corresponding table for the game looks like. I'm 'Johnny.' Actually, my first name is really János, but people call me Johnny in America. You're welcome to call me that, too. It's only with my surname that I'm a bit fussy: I won't let the Yankees take away my 'von,' no matter how republican they may be."

"I'm Oskar," says Morgenstern, in response to the mention of first names.

"Good, then I'll also write your first name," says von Neumann, now using the familiar "du" form in German to address Morgenstern. "I'll write down in the table how our points are shared out." Swiftly, he draws the table, fills it in, and shows it to his new friend:

		Oskar	
		matches	folds
Johnny	bluffs	1 0	0 1
	doesn't bluff	$^1/_2$ $^1/_2$	1 0

"Notice anything?" he asks proudly.

"It's the same table of numbers as before!"

"Correct. And it's obvious how this comes about. In the top left box, I have the king and am bluffing. If you call my bluff, I have to show the king: one point for you and none for me. If, on the other hand, you fold, then we're in the top right box—I don't need to show my card and my bluff has got me one point, and I write 0 for you. In the bottom right box, I am not bluffing, regardless of which card I draw. If it is the king, you immediately get the point. If it is the ace, you are due the point, because you have wisely folded and not matched me. Therefore, I have written 1 for you and 0 for me in this box. I'm also not bluffing in the bottom left box, regardless of my card. If it is the king, I concede defeat straightaway. If it is the ace, I raise the stake, you call and you lose. This is the playing strategy that I outlined at the beginning, with which after many rounds each of us has the same number of points. Therefore, I will give each of us half a point per round in this box."

"I understand. It's the same dilemma as with Holmes and Moriarty. If I assume that you are not bluffing, I will be clever enough to fold, since 'my' 1 on the right in the second line is greater than 'my' ½ on the left. But if you know that I am going to fold, then you are virtually forced to bluff, since 'your' 1 at the top in the second column is greater than 'your' 0 below. I can see that you can clearly describe with numbers the predicament in which the players find themselves. But the question is how one can free oneself from this predicament."

"In poker, it's obvious—I mustn't always bluff, but only sometimes. You mustn't always fold, but only sometimes. Let me illustrate this in a sketch." Von Neumann fishes out another sheet of paper and begins to draw.

First, he draws a straight line at a slight angle from the top left to the bottom right: "This straight line represents you matching me." He marks the line with "Oskar matches" and draws above it a parallel straight line that he marks with "Oskar folds": "This line represents you folding." He then draws a straight line at an angle from the bottom left to the top right, marking it with "Johnny doesn't bluff." Further down, he draws a parallel line and marks it with "Johnny bluffs." "These two lines represent my possible moves," explains von Neumann. "The four straight lines form a square . . ."

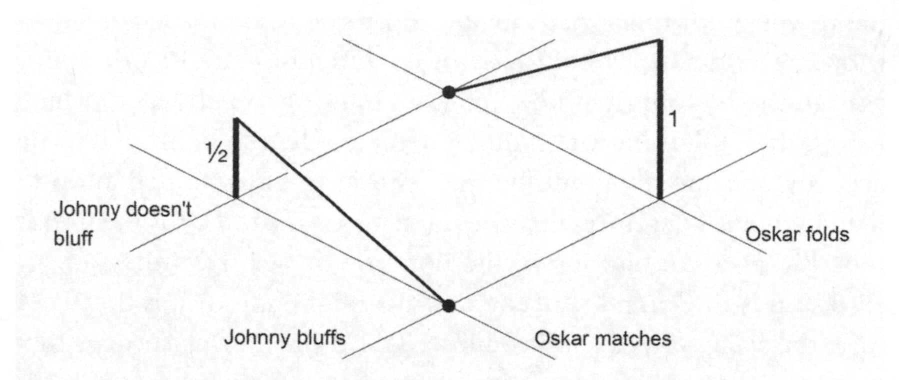

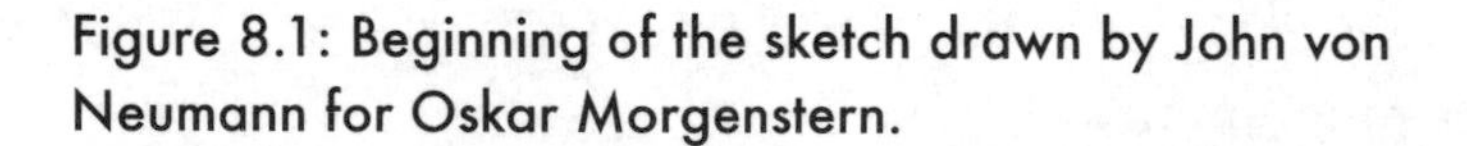
Figure 8.1: Beginning of the sketch drawn by John von Neumann for Oskar Morgenstern.

"Don't you mean a parallelogram?"

"No, no. You have to imagine that you are viewing this square at an angle from the side, since I am going to put in each corner of the square a perpendicular line showing what chances of winning I have in each corner. In the front corner, 'Johnny bluffs' meets 'Oskar matches': not good for me—I lose." Von Neumann draws a thick dot in this corner. "In the right corner, 'Johnny bluffs' meets 'Oskar folds': good for me—I win." Von Neumann draws in this corner a long, thick, perpendicular line and writes "1" next to it. "In the left corner, 'Johnny doesn't bluff' meets 'Oskar matches': there, it's fifty-fifty for us both." Von Neumann draws in this corner a thick, perpendicular line that is half as long as the previous one and writes "½" next to it. "And in the back corner, 'Johnny doesn't bluff' meets 'Oskar folds': again, bad for me and the point goes to you." Like in the front corner, von Neumann draws a thick dot in this corner. Then he continues his explanation:

"If I knew that you would match in every game, I would, geometrically speaking, keep to the front line running at an angle down to the bottom right. Above it, I'll draw a thick line that leads from the top point of the perpendicular line in the left corner, which is one-half high, to the thick dot at the front. The extent to which it is above the angled line gives me my chances of winning, according to how often I bluff or don't bluff. The more I bluff, the nearer I am to the thick dot at the front and the more foolishly I play, because my chances of winning are accordingly reduced toward zero. I am at my cleverest if I keep to the left on this straight line, at the top point of the perpendicular line, since my chances of winning are then at least one-half.

"The same is true vice versa if I keep to the back line running at an angle down to the bottom right, that is, I assume that you will always fold. Here, I will draw the thick line that leads from the thick dot at the back to the top point of the perpendicular line in the right corner, marked 1. Again, I can see what my chances of winning are according to how far above the angled line this thick line is. The more often I bluff here, the more I increase my chances, with winning guaranteed if I bluff in every single round."

"But since I will certainly not constantly match or constantly fold, I don't yet see how this illustration helps us," objects Morgenstern gently.

"That's right—we won't keep to the lines bordering the square. But each point in the square tells us with what frequency I bluff and you fold: the midpoint of the square, for example, represents that I bluff in half of all rounds and you fold in half of all rounds. And one point further back, near the back corner of the square, shows us that I bluff rather seldom and you fold rather a lot. Now imagine that a flexible and elastic rubber skin is stretched between the two thickly drawn lines." Von Neumann tries to illustrate this with a grid on the paper. "The perpendicular distance from this rubber skin to each of the points in the square tells me how great my chances of winning are. I can therefore use the rubber skin to calculate my chances of winning, if I know how often you fold and how often I bluff."

"The rubber skin has the approximate shape of a saddle."

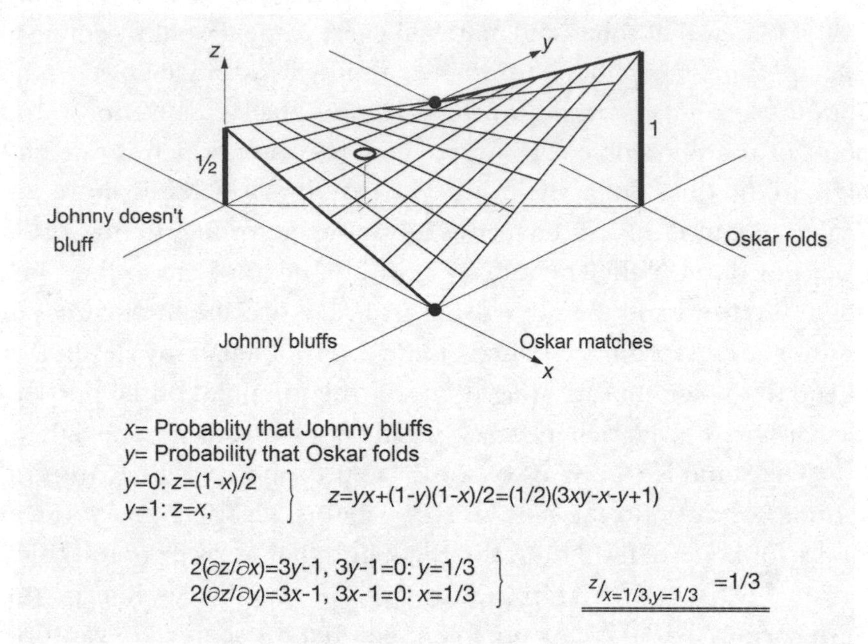

Figure 8.2: Sketch drawn by John von Neumann for Oskar Morgenstern. Underneath are the calculations used by von Neumann to determine the position and height of the saddle point on the "rubber skin."

"And the point on this saddle that a rider chooses as the point to place his weight, a point referred to by mathematicians as the 'saddle point,' leads us to the solution of the problem."

John von Neumann knows what he is talking about here. He understands all about riding—even though he only ever goes riding in a suit and tie. Very few people have ever seen him without a tie.

"The point of the square directly underneath the saddle point is the one that deserves our undivided attention. If I adjust the frequency of my bluffing according to this point, you can only move along the line passing through it, which runs at an angle from the bottom left to the top right, like 'Johnny bluffs' or 'Johnny doesn't bluff.' Regardless how often you decide to fold, you will never be able to reduce my chances of winning, which is the perpendicular line from the level of the square to the saddle point. This means that, exactly underneath the saddle point, I have maximized my minimum chance of winning. That's why I refer to the 'maximin rule.'"

"And finding the position of this saddle point is a routine calculation?"

"It only takes a few lines. Wait: if *x* denotes the probability that I bluff and *y* denotes the probability that you fold, then *x* and *y* are the coordinates of a point in the square. If I call the height of the rubber skin above this point *z*, then . . ." Oskar Morgenstern is speechless with admiration at how swiftly John von Neumann writes down one formula after the other, though he himself hasn't the faintest idea about them. Finally, von Neumann underlines the result twice and says quietly, speaking more to himself: "Of course, it has to be this way. For reasons of symmetry, the point of the square underneath the saddle point can be found on the diagonal line from the left-hand corner of the square to the right-hand one. And because the lengths of the perpendicular lines there have a ratio of one to two, this point must also be closer by a ratio of one to two to the left-hand corner than to the right-hand one." Then von Neumann looks up, sees Morgenstern's amazement and declares to him:

"The saddle point can be found where, on average, I bluff in one in three rounds and, on average, you fold in one in three rounds. There, I am at least guaranteed a one-third chance of winning. That's how far the saddle point lies above the surface of the square. A third is not much, but it is the maximum that I can achieve as a minimum gain."

"But it would be senseless if on a regular basis I were to fold in one round and then match in the next two. You would immediately see through such a strategy and would adapt your own game accordingly."

"Naturally, I should only bluff in one in three rounds 'on average,' and you should only fold in one in three rounds 'on average,' so that the frequencies come down to a third for both you and me. But distinct patterns in such decisions must be avoided like the plague. The wisest thing for us both would be to secretly throw a die before each round: if I throw a one or two, I undertake to bluff, otherwise I don't bluff. If you throw a one or two, you decide to fold, otherwise you match. In this way, we exploit the law of large numbers. That is the most sensible strategy for us both in this simple poker game."

"But what about Holmes? What should he do?"

"I could only advise him too to roll a die during the journey to Canterbury. If he throws a one or two, he should continue on to Dover. Otherwise, he should get off in Canterbury."

"Letting dice decide his fate, whether he lives or dies, seems rather reckless, don't you think?"

"Mathematics cannot offer poor Holmes anything better than that. At least he then knows that there is a probability of just over 33 percent that he will survive. But with regards to your question about the behavior of those operating in business—clients and companies—the maximin rule I have thought up is almost certainly of use. In business, you see, decisions come thick and fast, so it can be sensible to advise decision-makers how often they should make a positive decision and how often a negative one."

"We should write a book about this."

"Good idea—let's do that!"

"One more question: in your tables, you always entered two numbers into each box—the bottom left one for the player on the left, the top right one for the player at the top. Is that necessary at all? If I know what the bottom left number is, then I know what the top right one is too, since the numbers always add up to one. Would it not be sufficient just to write the number for the player on the left, for example, since the other one follows automatically?"

"A very good suggestion," says von Neumann appreciatively, "But I only wrote in both numbers today to help you understand better. I myself always do what you suggested. It's the same with all games—when I know what one player wins, I know at the same time what the other loses."

This last sentence pronounced by John von Neumann is true of the so-called "zero-sum games." That there can be other games than these was something that he simply couldn't imagine.

9

PLAYING WITH LIFE AND DEATH

BUDAPEST, 1908; PRINCETON, NEW JERSEY, BETWEEN 1929 AND 1957

"Why don't you join us in Vienna?"

It is at the beginning of the 1930s when John von Neumann, during one of his visits to Vienna, is asked this question by a member of the Vienna Circle. But he declines with thanks. "I tell you—Vienna's heyday will soon be over. On the other hand, the rich Americans are extremely interested in scientific progress and hardly anybody in Europe can keep up. My colleague Oswald Veblen has cofounded a magnificent Institute for Advanced Study. I have already settled in at Princeton—I've even changed my first name from Johann to John. America is the future. Take a look to the north: a barbaric rabble is gathering strength there, dragging Germany's name, Germany's culture, Germany's scholarship through the mud. I don't know if any state in Europe will be strong enough to stop those vandals. Indeed, I could ask you in return: Why don't you join us in the United States?"

"But you have old Austrian roots!"

"That is true," admits von Neumann. "To be more precise, I have Hungarian roots, but back then Hungary was indeed still part of the Danube monarchy. I can still remember clearly—I was ten years old at the time—when my father received his title of nobility. 'Neumann of Marghita'—how proud he was! It is in his honor—my parents were the best you can imagine—that I retain the 'von' in my name. But my

European roots have been cut. I am certain that I will fit in very well in my new homeland on the other side of the pond."

When the conversation is over, John von Neumann lets his mind wander back into the past. The image of his parents before his eyes, he recalls his family's large and beautiful apartment in Budapest, the splendid evening gatherings in the Neumann household, when he, a young lad of less than five, was introduced to the guests as a child prodigy by his father Max Neumann, a wealthy banker and royal state counsellor.

"Can you read already?" an elderly lady asks young János, as he was called then, whereupon Max Neumann tells a servant to fetch the Budapest telephone book. As soon as he gets the book—a relatively slim volume back then in 1908—Max tears a page out of it, hands it to his Jancsi, and instructs him to study the page in the closet and then return in five minutes. In the meantime, drinks are served and Max Neumann asks the lady to remain nearby. Then young János reappears with the page, which his father takes from him and hands to the lady, before inviting her to read out a name from the list.

"Darvas, Gábor," she says, and straightaway, the little stripling in his sailor's suit shoots back:

"Five-five-seven-nine-seven-six. Darvas, Gábor; he lives at number 15, Erzsébet királyne utcza. Please say a number this time."

"Three-two-eight-eight-four-one."

"That's the telephone number of Dávid, Gyula, who lives at number 4, Akaczfa utcza."

"That's incredible!" cries the lady in delight, as Max Neumann proudly sends his son back to the nurse.

Even before János attended elementary school, his father taught him Latin, closely followed by ancient Greek. The boy learned languages, living and dead, with remarkable ease. Above all, however, he had a stupendous talent for calculations. He could effortlessly divide an eight-digit number by a six-digit one in his head. The school to which his father sent him was the best in the city: the Fasori Evangelikus Gimnázium. The parallel with young Karl Menger, who attended the Döbling Gymnasium in Vienna at the same time, is striking: while Menger's fellow pupils Richard Kuhn and Wolfgang Pauli later won the Nobel Prize, von Neumann was schoolmates with Jenő Wigner, a physician who later in America called himself Eugene Wigner, and János Harsányi, an economist who, as an American citizen, was known as John Harsanyi, with both these men distinguishing themselves in later life. And, like old Carl Menger, von Neumann's father was horrified that his highly gifted son wished to eke out his existence with an unprofitable career like mathematics, particularly since Max von Neumann also arranged private tuition for his son outside of school, which meant that János had mastered differential and integral calculus by the age of fifteen, subjects that his private tutor Gábor Szegő, some nine years his senior, had only recently been exposed to at university. Szegő was so impressed by his young pupil's talent that he was moved to tears.

His father's instructions were to study chemistry, and János followed these obediently, graduating in chemical engineering at the Swiss Federal Institute of Technology in Zurich. At the same time, however, he continued with mathematics, at which he excelled, studying the subject at university in Budapest. He received his doctorate at the age of twenty-two, but had already, as a nineteen-year-old, published scientific papers on mathematics, laying the foundations for a career as one of the leading mathematicians of the twentieth century.

He wrote the article on the theory of games that he had mentioned to Oskar Morgenstern at the age of twenty-five. Here, he was entering uncharted territory, as in many of his papers on other mathematical topics. (It is true that, independently of Neumann, the French mathematician Émile Borel had anticipated a few of his ideas some years before, but he had only outlined them.) The geometrical interpretation of the simple poker game that John von Neumann presented to Oskar Morgenstern in the previous chapter is a good illustration of the novelty of von Neumann's discoveries.

In normal games of chance, for example those that Antoine Gombaud, whom we know as the Chevalier de Méré, was addicted to at the roulette table, the player can make various moves, but is in truth alone at the table. Whether the other players around him also bet a hundred livres on red when he does, or instead bet a thousand livres on black, makes no difference to him. Viewed in this way, pure games of chance are one-dimensional. The frequencies with which the player makes his moves in the individual rounds can be abstractly entered as points on a one-dimensional scale. A curve is stretched above this scale, and the distances between the scale points and the curve represent Gombaud's chances of winning.

In the parlor games that Neumann looked at, however, the player is not alone. He has a second player as an opponent, which makes the game two-dimensional. The simple poker game from the previous chapter is the best example of this: the frequency with which Johnny bluffs in the individual rounds is recorded on one scale as points, while that with which Oskar folds in each round is recorded on the other scale. The two scales enclose the square above which is stretched the rubber skin that tells us Johnny's chances of winning. This is the basis of John von Neumann's brilliant idea: the one-dimensional scale of the single gambler is replaced by the two-dimensional square of the two opponents in parlor games.

Realizing that the dimension increases with the number of players taking part in the game is no great discovery; the step from one to two is the real breakthrough. As an aside, even con games can be woven into this geometrical image—they are zero-dimensional games, since the conman's victim can make whatever moves he likes: he will get nowhere, his fate is sealed, and he is guaranteed to lose.

Shortly after the publication of his paper on the theory of parlor games, von Neumann decided to pack up and move to Princeton. This was a time when the National Socialist movement was still perceived by many not to be a danger, but von Neumann had a good nose for coming events and wanted to be on the safe side. For several of his colleagues who remained in Europe, it was a stroke of luck that the sophisticated and well-connected von Neumann had already established a foothold in the United States when the Nazis began to wreak havoc, much to the horror of the upstanding members of the academic community in Germany, and later in Austria and the other countries conquered by Hitler's troops, since he was in a position to smooth the path to America for them. One of the last to leave was Kurt Gödel, who only just managed to flee in time, at the wise insistence of his wife, Adele. They would never have reached the safe haven of Los Angeles and then found their new home in Princeton had von Neumann not insisted that Gödel join the Institute for Advanced Study.

At this research institute for the world's scientific elite, von Neumann would prove a mathematical theorem in one area of his subject in the mornings and another in a quite different area in the afternoons, and in practically every subdivision of mathematics he achieved significant results. He liked to spend his evenings and nights, however, in large social gatherings, and the soirees at the von Neumanns' Princeton home were legendary. Shortly before leaving for America, he had married Mariette Kövesi, for whose sake he had himself baptized as a Catholic—a purely formal act for him. With Mariette, and later with his second wife, Klara Dan, "Good Time Johnny" hosted the most opulent parties imaginable, with merriment and conviviality very much the order of the day. John von Neumann could converse on the most varied topics, with the history of antiquity a particular favorite—a professor of Byzantine studies once claimed that he could learn a lot from John von Neumann. He adored humorous stories and loved to joke around—on the most sophisticated level—ideally with his mathematics colleague Stanley Ulam; the two of them would outdo each other with wordplay in languages as diverse as English and Yiddish. Among the diners and at the bar, he always made himself out to be an inveterate drinker, though it was rumored that he simply used a few clever tricks to create this impression. And card games of all kinds, above all poker, were a constant feature—not for nothing had he compared Oskar Morgenstern's conundrum to a game of poker.

The joint book project on which Morgenstern and von Neumann

got to work after their first meeting made splendid progress. Two-player games, three-player games, four-player games—indeed games for any number of players were systematically explored and analyzed with the help of typical examples, all taking into consideration the fact that economic activity could be understood as a game. For the book was conceived for economists, not for professional gamblers, and its contents were not designed to show how to win games. Indeed, John von Neumann himself was a rather poor poker player. Rather, the two authors' aim was to analyze the structure of games and the structure of game strategies, similar to what John von Neumann had done with his simple game of poker.

In 1944, the book was finished: *Theory of Games and Economic Behavior*, written by John von Neumann and Oskar Morgenstern, was published by Princeton University Press. The critics almost fell over themselves to praise this weighty tome. Herbert A. Simon, the social scientist and later Nobel Prize laureate, encouraged "every social scientist who is convinced of the necessity for mathematizing social theory . . . to undertake the task of mastering the Theory of Games,"[1] while the probability specialist Arthur Herbert Copeland declared the book to be "one of the major scientific achievements of the first half of the twentieth century."[2] Leonid Hurwicz, an American economist and later Nobel Prize laureate who had emigrated from Europe, noted with foresight that "the techniques applied by the authors in tackling economic problems are of sufficient generality to be valid in political science, sociology, or even military strategy."[3] And after attesting to the book's "careful and rigorous spirit," the American economist Jacob Marschak, originally from Kiev, ended his review with the words, "Ten more such books and the progress of economics is assured."[4]

Despite all this glowing praise, most economists remained skeptical toward the book. This was nothing new for John von Neumann. His book on the mathematical foundations of quantum mechanics had not enjoyed the resounding success that might have been expected from the positive reviews, at least not among physicists, for whom the book was actually intended. He was all the more delighted, therefore, when the military began to take an interest in the book on game theory that he and Morgenstern had written. Leonid Hurwicz's words had turned out to be true: warfare is, from an abstract point of view, comparable to a zero-sum game. It is a "game" in which the "players" aim to make decisions that will bring the most benefit to themselves and the most harm to their opponents. That is why the maximin rule is of interest in this "game": how do we maximize the minimum gain that we can achieve, regardless of what the opponent does?

In 1948, the United States government founded the RAND Corporation in collaboration with private sponsors—RAND being an abbreviation of Research ANd Development. The RAND Corporation is the think tank for the American armed forces, and von Neumann's and Morgenstern's book was required reading for each of the corporation's employees. John von Neumann was flattered to be engaged as a scientific consultant for the RAND Corporation. Since his heavy involvement in the development of the atomic bomb during the Second World War, his ambition to do everything possible to strengthen the United States' military power had increased. The game theorists who acted in accordance with *Theory of Games and Economic Behavior* passed their first test in 1951 when the war in Korea broke out in earnest. They created a game table, not a 2 x 2 one like in the simple poker game example in the previous chapter, but one consisting of 3000 x 3000 boxes. They believed that they could use this to describe the war in terms of game theory. The optimum strategy was calculated on an ENIAC computer—the development of which, incidentally, was also thanks to John von Neumann. It is said that it was based on the resulting recommendation that President Harry Truman decided to relieve General Douglas MacArthur of his command.

Naturally, nobody can tell whether this decision really was the correct one. But President Dwight D. Eisenhower, Truman's successor and the Supreme Commander of the Allied Forces in Europe during the Second World War, decided in 1956 to reward John von Neumann for his services to the United States of America with the highest civilian honor, the Presidential Medal of Freedom established by Truman. Eagle-eyed observers will notice that the photograph that records the American president handing over the award certificate to John von Neumann shows that the recipient was confined to a wheelchair at the time. The smile on his face is strained, and not an expression of heartfelt joy, for John von Neumann was already aware that he was doomed to a premature death.

His curiosity about how the atomic bombs that he helped construct would explode seems to have been his downfall. He observed one of these nuclear tests on Bikini Atoll from recklessly close proximity and it can be assumed that the radioactive fallout caused the outbreak of a tumor that was later diagnosed as incurable.

"You must have gotten the results mixed up," he argued defiantly to begin with. "There must be something that can combat this hateful beast inside me." His desperation grew. "Why me, why now? I'm only fifty-three, I must live! What is this creature, this demon, doing inside my body? Stan, Edward!"—addressing his friends Stanley Ulam and Edward Teller, the so-called "father of the hydrogen bomb"—"Help me! Please say it's not true! Tell me that I shall live! I don't want to stop thinking, I won't, I won't, I won't . . ."

Teller later said that he had never seen anybody so desperate. "I think that von Neumann suffered more when his mind would no longer function, than I have ever seen any human being suffer."[5] Neighbors reported that the nights in the von Neumanns' house were full of screaming, groaning, and wailing. Terminally ill, von Neumann simply could not accept that his prodigious intellect, of which the Nobel Prize–winning physicist Hans Bethe once said that it was evidence that superhuman intelligence must exist, should be consigned to extinction.

When he was transferred to the Walter Reed Army Medical Center, military guards were stationed before his room day and night. The military commanders feared that, as an employee of the RAND Corporation, he might reveal top-secret information in his delirium, but John von Neumann merely lamented his own fate.

Shortly before his death, however, something wondrous occurred that nobody who had known John von Neumann in his heyday could explain: feeling his mental capabilities fading away (the cancer had also attacked his brain), he begged that Father Anselm Strittmatter, a Catholic priest of his acquaintance from the Benedictine order and a highly learned man, should come to see him.

"Iudex ergo cum sedebit, quidquid latet, apparebit, nil inultum remanebit. Quid sum miser tunc dicturus, quem patronum rogaturus, cum vix iustus sit securus?" von Neumann moans, and makes a tortured attempt to produce the melody from Mozart's Requiem. The English translation of these lines from the *Dies Irae* is, "When therefore the Judge will sit, whatever lies hidden will appear: nothing will remain unpunished. What then will I, poor wretch, say? Which patron will I entreat, when even the just may hardly be sure?"

"Salva me, salva me, fons pietatis," (Save me, save me, O font of mercy) answers the priest as he enters, intoning Mozart's magnificent version of this passage. The soldiers on guard haven't the faintest idea, of course, what the two men then talk about, since Strittmatter and von Neumann continue to speak Latin until the priest leaves.

"To the very end, Johnny had no interest in religion—he simply wanted to have somebody there with whom he could speak Latin and ancient Greek," says one of his close friends flippantly at the funeral.

"You may be right," whispers his neighbor. "Father Strittmatter didn't tell me what they spoke about. But he did confirm one thing—Johnny didn't die in peace. Right until the bitter end, he trembled before death."

"And yet it is odd," adds another mourner, who has overheard the conversation. "He did get the priest to give him the Sacrament of Anointing of the Sick. And the last time I visited him, a few weeks ago, he spoke to me, almost in tears, about Pascal's Wager."

"Pascal's Wager?" asks the first man.

"It's a kind of game dreamt up by Pascal," responds the other. "It's all about the existence of God."

"Typical Johnny," murmurs the first man, as he takes a shovel and lets earth trickle down onto the coffin in the open grave. "The old fellow only ever had games on his mind."

10

PLAYING WITH CHICKENS AND LIONS

PRINCETON, NEW JERSEY, 1949

"This man is a mathematical genius."

The letterhead of the Carnegie Institute of Technology is emblazoned at the top of the page, while at the bottom is the signature of Richard Duffin, a professor of electrodynamics there who has made a name for himself with his research into electrical circuits. Apart from this single six-word line, the page is blank. Emil Artin, the famous Vienna-born mathematician who emigrated from Hamburg, was briefly a colleague of Karl Menger's at the University of Notre Dame, and now teaches at Princeton University, turns the sheet over. But the other side is empty. The only thing written on the paper is this single sentence: "This man is a mathematical genius."

Before Artin stands the "mathematical genius," a tall, confident young man with a slightly arrogant gaze. Artin attempts to defuse the crackling tension between the two of them. "Well, Mr. Nash," he says, "you may be a genius in Carnegie for this physicist Duffin. But you know what? Here in Princeton, what he has written in this funny little reference doesn't count for much. We're all mathematical geniuses here." A tense silence follows for a few moments. "Which lectures are you going to attend this year?"

"I have no intention of going to lectures," answers Nash with supreme self-confidence. "I work on mathematics by myself and merely want to exchange ideas with a few professors from whom I expect to get useful information."

"I see," replies Artin, and the icy undertone is unmistakable in his voice. He takes up the book he was reading when interrupted by John Nash's entrance, and the young man realizes that the short interview has come to an end—and that Emil Artin is by no means a fan of his.

Solomon Lefschetz, who had held a chair in mathematics at Princeton University since 1925, invited John Nash to continue his math studies with him at Princeton, rather than at faraway Harvard. It may well be that Lefschetz was so impressed by Duffin's reference because, like Duffin, he had actually wanted to be an engineer. Alas, both of his hands were torn from his body in an accident in 1907 and he was forced to change his career path and become an academic. He turned to mathematics and developed into a master of his special field, topology. After spells in Nebraska and Kansas, he ended up in Princeton in his forties and became a ubiquitous presence at the university. His energetic nature, brawny appearance, pugnacious language, and fiery delivery were a striking contrast to the cool reserve and aristocratic elegance of Emil Artin. When Solomon Lefschetz entered the lecture hall, a chosen student would place a long piece of chalk between two fingers of his black-gloved prosthetic hand, and this stub of chalk would remain there until the late afternoon, while he spoke forcefully to students and colleagues alike with his powerful voice. His tireless commitment to the advancement of mathematical research at his institute was a deciding factor in the University of Princeton gaining such an excellent reputation for mathematics, and it was no wonder that the Institute for Advanced Study decided to pitch its tent nearby. When Solomon Lefschetz insisted that a young talent like John Nash should study at Princeton, nobody dared to contradict him, regardless of the young man's airs and graces. Not even the skeptical Artin.

With the reference "This man is a mathematical genius" in his hand, Nash introduces himself not only to the university professors but also to the renowned researchers at the Institute for Advanced Study. With the world-famous Robert Oppenheimer, the director of the Institute and the physicist who organized the construction of the first atomic bomb, he does not experience the same disappointment as with Artin, but their talk, though polite, is noncommittal. Oppenheimer clearly has other things to worry about than enrolling newcomers, though he can at least organize an appointment with the legendary Albert Einstein. But when Nash calls on Einstein at his house in Mercer Street and spends almost an hour enthusing about his ideas for how the theory of gravity could be rewritten—his aim being to impress the famous old professor—Einstein merely listens benevolently to him, sucking on his pipe all the while, and dismisses him with a single sentence: "Young man, you still have plenty to learn about physics."[1]

John Nash doesn't let this put him off, however. In keeping with his stated intention, he attends hardly any of the lectures and none of the seminars at the university. He even considers many books to be a waste of time. His fellow students generally find him sitting and musing in the communal rooms, and even when several dozen of them gather in a hall John Nash always stands out as a bit of an oddball due to his aloof, withdrawn, and complacent nature. He will suddenly speak to his neighbor, somebody he has never seen before, in order to recount a flash of inspiration he has just had. Without waiting for the other's answer, he strides to a table and makes the strangest notes on one of the pieces of paper lying there. Then he remains seated at the table, staring absentmindedly into the distance, his motionlessness only interrupted when he notices a few students at the next table passing the time with some tricky game of intelligence. John Nash sidles up to them and at first follows the students' moves in silence. But after a short while, he begins to make comments, generally mocking, which naturally doesn't win him any friends. But this doesn't bother him. He isn't at Princeton to get to know kindred spirits—he is at Princeton to become one of the best mathematicians in the world.

Those who have already achieved this goal always meet up for tea in Fuld Hall at three o'clock, and students are permitted to join them when the luminaries are casually discussing mathematical questions. John Nash can be found here without fail. He collects questions that remain unanswered and later strives to illuminate them from every possible angle on countless sheets of paper, which almost always end up in the wastepaper basket. He knows full well that his approach is an unconventional one, since he scorns the text books in which the traditional ways are described. The fact that his methods soon lead him to a dead end does not bother him in the slightest. The professors meet again the next day, and new problems are put forward for discussion.

It may have been von Neumann himself who one day happened to speak of the game theory that he had published in a thick book together with Morgenstern. John Nash hears of this, flicks through a few pages of the renowned book, and begins to formulate his own thoughts on games. Suddenly, he is convinced that he has discovered something new.

"Mr. Nash, what can I do for you?" Albert Tucker is one of the professors at Princeton whom the students can always approach in the knowledge that they will never be turned away.

"Professor Tucker, you read about last week's accident on the West Coast, didn't you—two teenagers in stolen cars crashed into each other and neither survived."

"Yes, I vaguely remember, but I'm not sure . . ."

Nash unceremoniously interrupts Tucker. "I believe the two of them were playing chicken."

"Chicken?"

"Yes. For teenagers, 'chicken' means 'coward.' When you play chicken, you want to show your opponent that you're not a coward. The two of you steal two old bangers and, on some deserted country road, drive at top speed straight toward each other. The first one to swerve away to the right is then the chicken and the other, who was brave enough to drive straight on, is the lion, the hero."

Tucker is still staring uncomprehendingly at the wildly gesticulating young man, but Nash is only just getting into his stride. "The game of chicken isn't mentioned in the book of game theory by Morgenstern and von Neumann," he announces with satisfaction, "since it isn't a zero-sum game." He blithely takes a piece of paper and one of the pencils from Tucker's desk. "I thought of it like this: let's call the two boys Jim and Buzz. Each of them has two alternatives—continue driving or swerve away. In my view, that results in the following table:

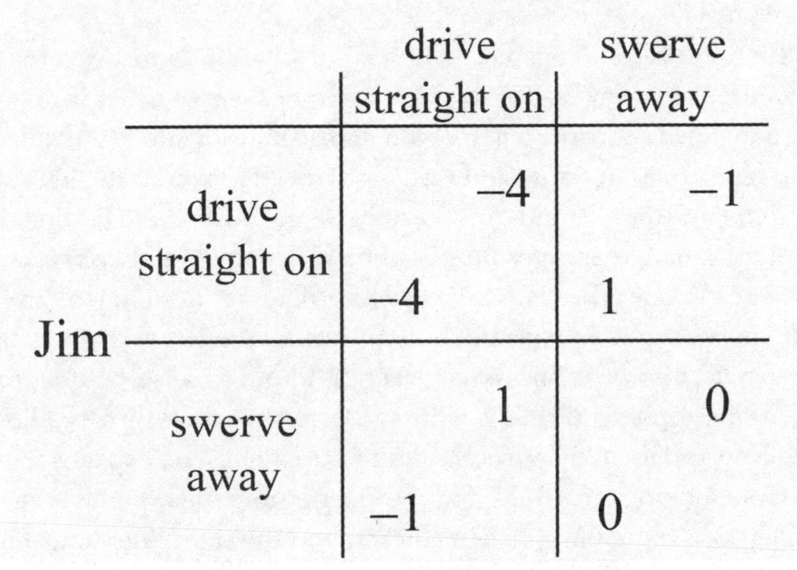

		Buzz	
		drive straight on	swerve away
Jim	drive straight on	–4 –4	–1 1
	swerve away	1 –1	0 0

"The two zeros at the bottom are obvious: if both of them swerve away, neither of them can show off to the other. Neither of them has won, but equally neither of them has lost. If both of them continue driving straight on, the cars crash into each other, which in itself isn't necessarily too bad a thing, since they're stolen. But the two boys risk being seriously hurt. That's why I put minus four in the top left box—minus, because you're talking about the opposite of winning."

"But why minus four and not some other negative figure?"

"It's all the same—you could just as easily put minus ten," is Nash's curt response, but then he suddenly remembers the rules of good behavior, rules that do not come naturally from his inner character, but which he has taught himself, like others might learn vocabulary in a foreign language. "Mr. Tucker, sir, take a look at the other two boxes. In the top right is what the players get if Jim continues driving and Buzz swerves away. In such a case, Jim gets plus one and Buzz minus one. and in the bottom left is the opposite case—Jim swerves away and so is penalized with minus one, since he is the chicken. But Buzz, who continues driving straight on, has won the game and gets plus one."

Tucker's interest has been fully aroused. "I see exactly what you're saying."

"It's not a zero-sum game," says Nash exultantly, "and I've already analyzed it." He beams at Tucker, takes another sheet of paper from the desk, and draws a square on it. He extends the bottom side of the square to the right, adds an arrow, and writes *x* there. He extends the left side of the square upward, adds another arrow and writes *y*. "The bottom side of the square represents Buzz as a chicken, that is, he always swerves away. The top side represents Buzz as a lion, always continuing to drive." As he speaks, Nash extends the bottom line to the left and writes BC next to it, to show that Buzz is a chicken. Without a pause, he does the same to the top line, this time writing BL, to show that Buzz is a lion. "The same applies to the vertical sides of the square," he explains as he draws, not looking up at Tucker and concentrating solely on his sketch. "The left side represents Jim as a chicken and the right side shows him stubbornly continuing to drive." Again, he extends the lines and adds JC on the left and JL on the right. Then he writes a capital B at the top right of the page, circles it, and commences with his lecture again.

"It's not a zero-sum game," he repeats, obviously infinitely pleased by this fact, "so I have to view the situation for each of the players individually. I'll start off with Buzz. If he knows that Jim will most likely swerve away—we're in the left-hand side of the square in the diagram now—he has a clear strategy: keep driving." Nash uses his pencil to make the top side of the square thicker at the left.

"However, if he knows that Jim is most likely to play the lion, that puts us on the right hand side of the square. In that case, it makes sense for him to swerve away." Now Nash presses hard with the pencil on the right side of the bottom side of the square. "Somewhere in the middle, Buzz's strategy switches from driving straight on to swerving away, and I immediately worked out where that will be." Nash completes his diagram, his pencil drawing a short, thick line from the top left corner of the square to the right, falling abruptly and vertically to the *x* axis and then turning horizontally along the bottom line to the right-hand corner of the square, thus creating a hook shape. "This hook," he explains, "is drawn in such a way that the left-hand rectangle it borders off is exactly a quarter of the square."

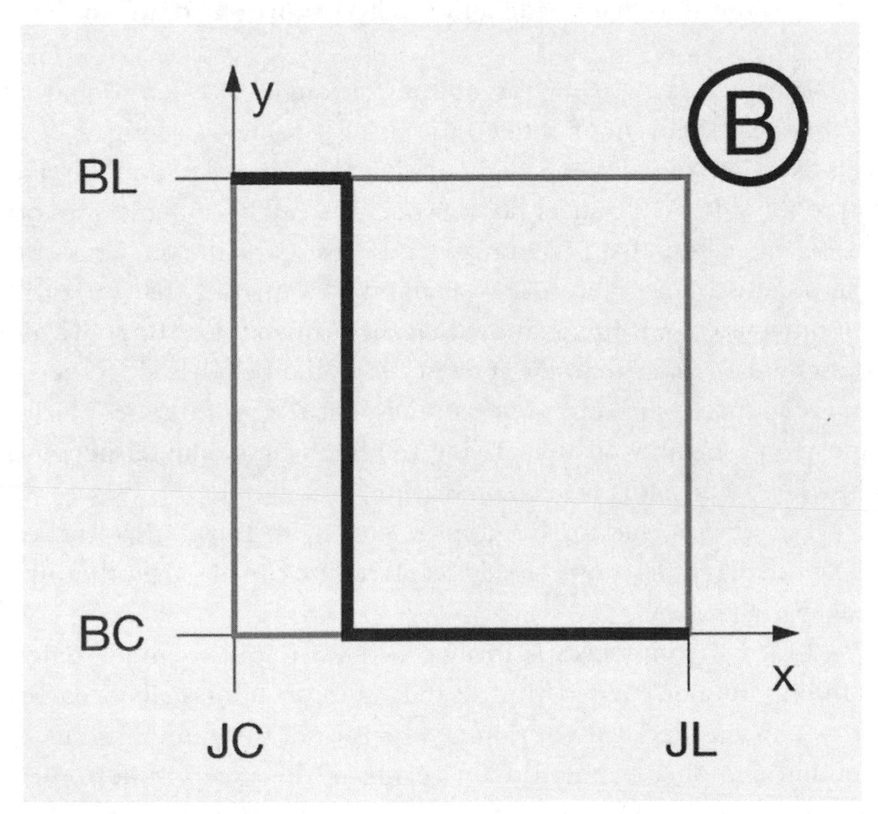

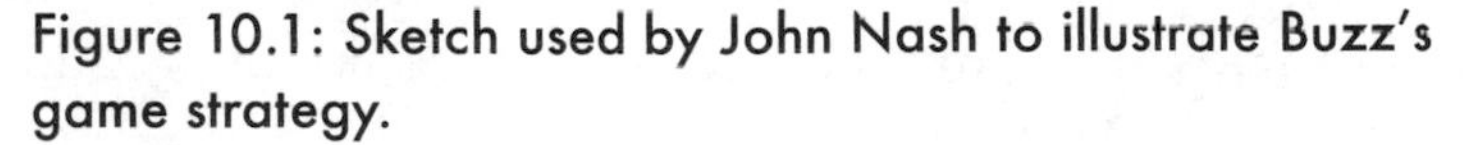
Figure 10.1: Sketch used by John Nash to illustrate Buzz's game strategy.

"The quarter must come from the fact that, in the top left box of the square, you have put minus four as a 'punishment' for both of them carrying straight on," says Tucker. "If you put minus ten there, the vertical line would be further to the left and the separated rectangle would take up a tenth of the entire square."

"Exactly!" says Nash, delighted that Tucker can obviously follow his thinking, but before he can continue talking, it is Tucker's turn to interrupt the younger man.

"If you look at the game from Jim's perspective, your drawing will be like this," he says, and takes a piece of paper himself, sketching a similar diagram to Nash's, but with a thick hook shape leading down for quite a lot of the left-hand side of the square, then horizontally across to the right-hand side and finally a short way down to the *x* axis.

"Exactly, exactly! Now the bottom rectangle takes up a quarter of the square. This hook reflects Jim being sensible—as long as Buzz acts like a lion and drives straight on, it makes sense for Jim to swerve off to the side. At a critical point, where the hook suddenly goes off to the right, Jim changes strategy and drives straight on. This critical point is where there is a probability of a quarter that Buzz will continue driving. If Jim assumes that the probability of Buzz driving straight on is more than 25 percent, he would be advised to swerve away. On the other hand, if he assumes that Buzz is rather cowardly, and the probability of him playing the lion is less than 25 percent, then he should actually keep on driving."

"But what is true for Jim applies equally to Buzz," says Tucker. "The two diagrams should really be placed on top of each other like transparent slides."

"Exactly!" John Nash is thrilled to have found such an understanding listener. "When you do that, the two hook-shaped curves cross one another at three points, which I call the game's points of equilibrium: once in the top left corner of the square, where Buzz stubbornly keeps on driving and Jim keeps on swerving away, then diametrically opposite this in the bottom right corner, where Buzz is the constant chicken and Jim always plays the lion. Lastly—and this strikes me as the most fascinating point of equilibrium here—they cross at the point where the vertical and horizontal dividing lines meet. Here, there is a 25 percent probability that both Buzz and Jim will continue driving straight on. At all three points of equilibrium, there isn't a good reason for either Buzz or Jim to renounce the strategy that applies to each point of equilibrium."

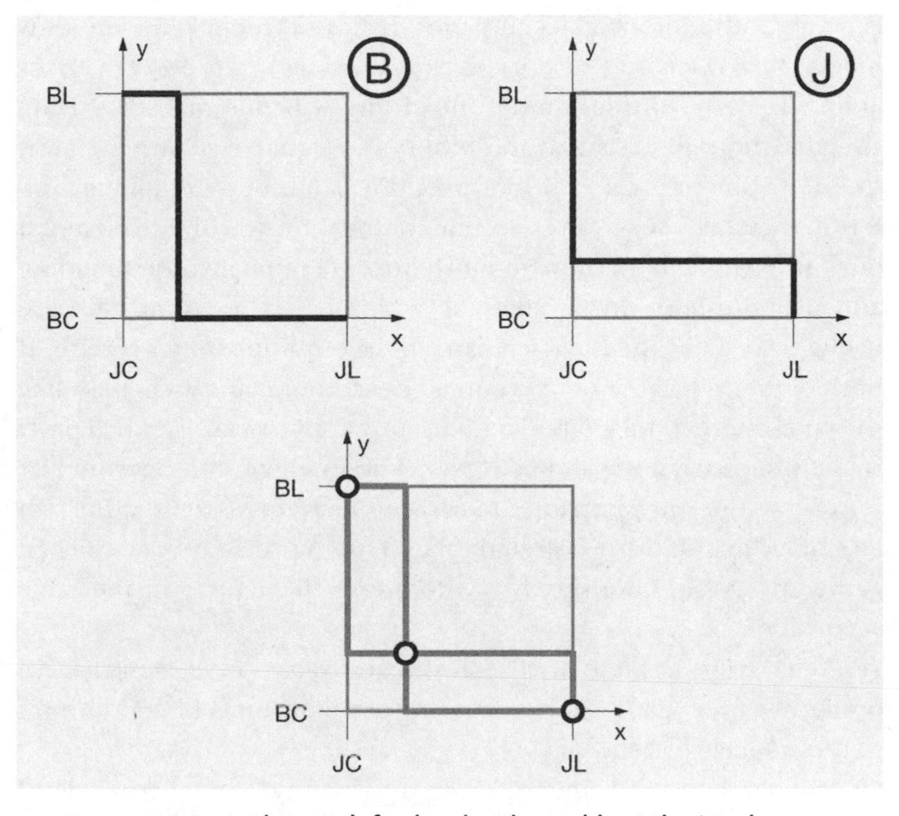

Figure 10.2: At the top left, the sketch used by John Nash to illustrate Buzz's game strategy; at the top right, the sketch used by Albert Tucker to illustrate Jim's game strategy. If one places the two sketches on top of one another, this gives the bottom diagram, where the three points of the Nash equilibrium in the game of chicken can be seen.

"Very impressive indeed," says Tucker, as Nash comes to the end of his lecture. "You should go to John von Neumann and show it to him."

"I've already been to see him," replies Nash, and adds with complete indifference, "He threw me out of his room. 'That's trivial and nothing new,' he snapped, and said it was just another variety of a long-known fixed-point theorem. But before I believe him, I wanted to come to you, Professor Tucker, and ask you if it really is nothing new."

"I'm certain that it is new. And it sounds highly interesting. We can forget the idiotic chicken game. It doesn't help Jim if he only meets Buzz once and decides to swerve away at 25 percent probability. He can't divide himself into four, with one quarter driving straight on at all costs and the other three quarters swerving away to safety. But you can also interpret the game in a completely different way. Let's say two restaurant chains want to run a canteen at our university. If both of them move into the same institute building and play the lion, they'll compete with each other until both are brought to their knees, which can't be the point of the exercise. If both of them decide not to set up a branch here, and so behave like the chicken, then they'll both end up unhappy, because a third party is likely to come along and take over. The intelligent thing would be for one of the two companies to run our canteen and the other not to—but which of the two should it be? That is precisely your chicken game, Mr. Nash. One simply has to adapt the figures in the table accordingly."

"That's no problem at all. On the contrary, I'm certain that, in practically every game, one would find similar points of equilibrium to those I have just shown you."

"Good, then think about it. You can write a doctoral thesis on it, and I would be happy to be your supervisor."

Nash doesn't acknowledge this offer. "The saddle point that von Neumann obtained with his maximin rule is merely a trivial example of a point of equilibrium as identified by me," he says, speaking more to himself than to his future doctoral supervisor. "Seen in this light, my theory goes far beyond John von Neumann's. But I have no intention of telling him about it . . ." He gathers the papers together and leaves Tucker's room without so much as a farewell.

Tucker is still gazing out of the window as Nash strides along the path with his head held high and his back straight. An odd fellow, the professor thinks to himself, highly gifted without a doubt, but also mentally unstable. He has no idea how accurate this assessment will prove to be. When John Nash's dissertation, amounting to less than thirty pages, was published, the RAND Corporation expressed an interest in the young genius. The Cold War had broken out between the United States and the Soviet Union, and neither of the two powers wanted to be the chicken while the other presented itself as the lion. Nash was enlisted to devise game theory scenarios for the RAND Corporation. Rumor has it that the RAND Corporation's advice to President Kennedy during the Cuban Crisis, which brought the world to the very brink of war, played a decisive role in defusing the situation. What is certain, however, is that John Nash's fragile mental state became even more vulnerable as a result of his work with political and military strategies and culminated in the onset of paranoid schizophrenia. He was plagued by constant delusions, sensory processing disorders, and acoustic hallucinations, making further work as a mathematician inconceivable. It was only thanks to his wife, Alicia, whom he had married shortly before becoming ill, and to his former fellow student John Milnor that he was allowed to remain in Princeton in spite of his peculiar behavior. The students would see him walking by himself in the campus, occasionally giving strange messages devoid of sense to people completely unknown to him, and he was dubbed "the Phantom of Princeton." But thirty years after the onset of his illness, Alicia Nash and a few other people who had contact with him noticed that he had begun to speak sense again and even started tackling mathematical problems once more. In 1994, the risk was taken to bestow on him, together with Reinhard Selten and John Harsanyi, the Nobel Prize for their joint achievements in the field of game theory—a "risk" because it wasn't certain if he would observe the usual decorum during the ceremony. But he was not permitted to give an acceptance speech such as Russell Crowe did as Nash in the film *A Beautiful Mind.*

As Tucker stands at the window, watching his newly acquired doctoral candidate John Nash disappear behind the hills of the campus, he realizes why economists have till now shown such little interest in the game theory conceived by Oskar Morgenstern and John von Neumann. In the authors' famous book, the two men assume that the players of the games they examine are interested in causing the greatest possible damage to their opponents, since that is the only way to gain the greatest possible profit for oneself in a zero-sum game. But if one looks at games that are not zero-sum games, the players are solely interested in their own gain, which does not necessarily have to equate to the damage of the other party. A player who maximizes his own payout doesn't have to thereby minimize his opponent's.

It is this aspect, conceived by John Nash, that finally brings to fruition the promise made by John von Neumann and Oskar Morgenstern in the title of their book: game theory explains economic behavior. Perhaps it even explains the behavior of mankind. Albert Tucker turns these thoughts over in his mind as he packs his suitcase. He has been invited to Stanford on the West Coast of the United States for a seminar of lectures. He is to give talks to students of psychology about possible uses of mathematics in their field.

11

PLAYING WITH PRISONERS

STANFORD, NEAR PALO ALTO, CALIFORNIA, 1949

"Over there is the Alcatraz Federal Penitentiary."

Before carrying on to Palo Alto and thence to Stanford, Albert Tucker spends a few days in San Francisco: Nob Hill and the cable cars, Fisherman's Wharf and the old Mission San Francisco de Asís, along with the beautiful city's many other attractions, all demand to be seen. He also joins a guided tour to the Golden Gate Bridge, at the time of its construction the longest suspension bridge in the world. Just as the mathematician in him recognizes the bridge's catenary curve, the so-called "cosinus hyperbolicus," he hears the eloquent guide pointing out Alcatraz Island over his megaphone. There is to be found America's most notorious top-security prison, one it is impossible to escape from.

Nobody in the tour group can understand why Tucker suddenly begins to smile when he looks at Alcatraz Island. Is somebody incarcerated there who once caused him harm? Is he in love with a woman who works in the security service there? The reality is nothing of the sort—Tucker is smiling because he has just recalled a story told to him by Melvin Dresher, a mathematician working at RAND, about two prisoners. He decides to tell his psychology students at Stanford his own version of this story. He has to take into account the fact that his audience knows little about mathematics, but he is sure that they can handle a table of the kind John Nash had showed him.

At the university, Albert Tucker begins his lecture in the packed lecture hall. "Ladies and gentlemen, you know the provision in American law that an accomplice witness to a crime can be granted a reduced sentence—indeed sometimes even an exemption—for testifying against a co-defendant. I'd now like you to imagine the following situation: two gangsters are arrested by the police. I don't know any current criminal names and so will simply call the first one 'Al' and the second one 'Capone.'"

"Why doesn't he call them 'Al' and 'Bert'?" whispers one student to his neighbor, who has to stifle her giggles so as not to interrupt Albert Tucker.

"The police suspect the two gangsters of having robbed a bank. For bank robbery, the judge can hand out a jail sentence of between six and eight years—six years, however, only if the defendant confesses to the crime from the start. There is no evidence to back up the police's suspicions in this case, however. The only crime that can be proved is illegal possession of firearms, which carries a sentence of two years in prison.

"However, the district attorney has the following idea about how to convict the two men of bank robbery: he has them put in two separate cells so that they cannot communicate with each other. Then he goes into Capone's cell, introduces himself, and makes him the following offer: 'Capone, you and I, we both know that you robbed the bank. But you also know that we don't have any proof of this. If you deny having robbed the bank, therefore, you can expect a sentence of just two years for illegal possession of firearms. But I warn you: I'm next going to go and see Al in his cell and am going to try to persuade him to confess. If Al confesses to the bank robbery, then you're in deep trouble and will spend eight years in prison because you denied the crime. He, on the other hand, goes free because he confesses and testifies against you. Naturally, you also have the opportunity to go free if you confess to the bank robbery and Al keeps his mouth shut. Think about it. I'll be back tomorrow. The trial is the day after tomorrow.'

"The cell door is opened, the district attorney bids Capone farewell, and, after the door has been locked again, Capone begins to brood on the matter, since he knows that the district attorney will now make Al the same offer in his cell."

"But shouldn't Al be afraid that, if he goes free as a witness for the prosecution, he will be killed the following day by somebody from Capone's band?" calls out a student in the back row.

"That's true, but we don't need to take that sort of thing into account. As you can see, I am thinking like a mathematician here and trying to leave out unnecessary complications, like the fact that other prisoners could pass secret messages between Al and Capone. I am ignoring all of these possibilities. For me, they are like background noise that only distracts me from my true goal, which is to illustrate the two crooks' situation to you as clearly as possible. To this end, let me draw the following sketch on the board."

Tucker takes a piece of chalk and sketches clearly on the board the following table:

		Al	
		keep his mouth shut	come clean
Capone	keep his mouth shut	−2 −2	0 −8
	come clean	−8 0	−6 −6

"As you can see, I have written the names of the two rogues on the left and at the top, with the two alternatives they face by their names. Both of them can either keep their mouths shut or come clean. In the boxes, I have written what each decision means for them. If both keep their mouths shut, that puts me in the top left box: they both get two years for illegal possession of firearms. If they both come clean, that puts me in the bottom right box: the district attorney doesn't need an accomplice witness for the prosecution anymore; they have both confessed and so must both spend six years behind bars. You will notice, by the way, that I have written all the numbers apart from zero with a negative sign, since these years will be taken away from the free lives of the defendants.

"Now let us look at the top right box. It depicts the situation whereby Capone, as a rogue of honor, keeps his mouth shut, while Al blows the whistle on him and goes free as a reward, leaving Capone to stew in prison for a full eight years. The bottom left box shows the same situation, only with the roles reversed.

"Imagine, ladies and gentlemen, that you are in Capone's place and are weighing up what to do. Which of you—please put your hands up—would keep your mouths shut?"

Gradually, hands are raised in the lecture hall, until after a few seconds a sizeable majority of those present have their hands in the air. Tucker smiles around the hall.

"Very nice. I assume that everybody else would come clean."

"Not me," objects a young lady on the left in the front row. "I abstain—I'm simply curious to see what Capone would really do."

"I assume that he's a cunning fellow," answers Tucker, his eyes on the student, "and draws, as we have done, this table. First, he looks at the two left-hand boxes. Which is the better option for him, assuming that Al doesn't confess—two years behind bars if he, Capone, also keeps quiet, or freedom if he comes clean? In this case, it's more sensible to confess. Then he looks at the two boxes on the right. Which is the better option for him if Al confesses—eight years in jail if he keeps his mouth shut or just six years if he also confesses? Naturally, he would also confess in such a case. In any event, the most sensible course of action is to confess."

"But that's absurd!" protests a student sitting in the middle of the front row. "The most sensible thing is for them both to keep their mouths shut."

"Precisely," agrees the young man behind him. "Because if you count up the years in prison of the two men in each of the boxes, you get a total of twelve years behind bars in the bottom right box, with both of them coming clean, while in the top right and bottom left you get only eight years, and in the bottom left, where both of them keep quiet, it's only four years. The two of them just have to keep their mouths shut!"

"Don't forget that the district attorney ensures that they cannot communicate with each other," counters Tucker. "Even if Capone is aware that honor among thieves dictates that he should keep quiet, he can't be sure that Al will stick to this unwritten rule. He cannot sleep all night, for fear of the eight years that he must spend in jail if he keeps quiet and Al confesses. Forget honor among thieves—when the district attorney comes into his cell the next day, he will confess, and thereby bring six years in prison upon himself, because Al, faced with the same dilemma, will also confess. That's why I call this game played with the two criminals by the district attorney the prisoners' dilemma. The prisoners could save themselves years in prison, but they won't do it."

"I don't think it's a dilemma at all," says the young lady on the left. "I'm on the side of the district attorney, who wants to take no chances and so uses this method to ensure that wrongdoers are justly punished in accordance with the laws of the land. For me, a dilemma is something quite different."

How right she is, thinks Albert Tucker, when he hears her words. His prisoners' dilemma is much more simply structured than the game of chicken that John Nash had told him about before his journey to California. In the game of chicken, there are three points of equilibrium, whereas in the prisoners' dilemma, there is just one: the corner of the square where both players come clean. Yet, although this curious game is so simple in principle, Morgenstern and John von Neumann have overlooked it. But Tucker has no wish to say all this in front of the budding psychologists, since it would be a departure from his actual subject, and so his answer to the student and the rest of the audience is quite different.

"I too am on the side of the law, like you. Consider the prisoners' dilemma merely as an example of the games I would like to introduce you to. As students of psychology, you surely know how easily people can be seduced into playing games. I'd therefore like to present a second game now. It has more to do with business than with criminality. You will even recognize that it reveals a great deal about the behavior of businessmen and companies. It is not only a game—it is also applied, useful mathematics. Let me first choose two players from among you." He turns to the young lady in the first row. "How about you—perhaps you would be so kind as to tell me your name?"

"Ann."

"Good, and who would like to play with Ann—perhaps you, young man? What is your name?"

"Bob."

"Excellent. I'd like to ask the other ladies and gentlemen to carefully follow the game that Ann and Bob are going to play with me.

"I, as the game master, am a kind of profitable company with whom people invest their money for a significant return. I come to each of you two, Ann and Bob, and ask you whether you are prepared to invest 80,000 dollars in my company. You can invest this large sum of money or you can decline. I promise you—with the guarantee of zero risk—the following: the money that the two of you invest in the company delivers a 50 percent profit after one year. I then divide all of the money, that is, your investments and the profit, equally between the two of you, Ann and Bob. That is the game.

"It would seem that such a game provides a win-win situation. Let's imagine that Ann and Bob both invest 80,000 dollars—a huge amount. I increase the entire sum—160,000 dollars—by 50 percent, so that after one year I have 240,000 dollars to pay out in equal portions to the investors. Ann receives 120,000 dollars and so does Bob, meaning both of them get a fat profit of 40,000 dollars.

"That's how it looks in theory. But now let me actually ask you, Bob. Will you invest 80,000 dollars, or do you decline this investment?"

"Of course—I couldn't say no to such a deal," answers Bob.

"And what about you, Ann?" As Albert Tucker poses the question, he shakes his head from side to side to make it clear to her that he wants her to refuse.

Obediently, she answers, "I won't invest."

"Now let us see how much money I have to increase by 50 percent," says Tucker to the audience. "80,000 dollars, so let's add 40,000 to that. Now I divide it up equally: 60,000 dollars for Ann"—Tucker simulates handing over the money to her—"which is 60,000 dollars of pure profit for you—more than what I said before. And 60,000 dollars for Bob"—Bob is rather startled to hear Tucker say this as he pretends to hand over the money—"which is 20,000 dollars less than you invested. A significant loss. How about if we play the game again? What will you do now, Bob—will you invest 80,000 dollars again?"

"I'm not mad!" says Bob resoundingly. "One loss is enough. Of course I will decline."

"What about you, Ann?"

"I will of course also decline."

Tucker is visibly pleased by their answers. "If I only bring in zero dollars of investment, the increase by 50 percent isn't of much use to me. If you add 50 percent of zero to zero, you still get zero. So my two players each get half of zero, which may not be less than zero, but it isn't more, either."

He takes up his chalk again and draws on the board next to the table for the prisoners' dilemma a new table for the game, with Ann and Bob as the players. "I won't write all the zeros for the money in the table," he comments. "I'll just write how many lots of 10,000 dollars the players record as profit or loss."

"Do you notice anything?" he asks loudly.

There are a few seconds of tense silence, and then a student in the last row speaks up: "If you take away six from every number in your table, you get the table from the prisoners' dilemma."

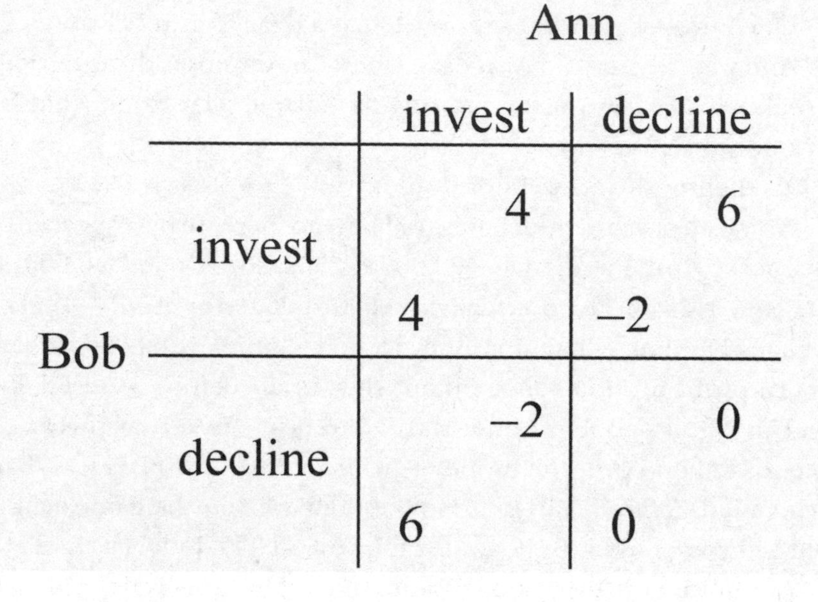

		Ann	
		invest	decline
Bob	invest	4 / 4	6 / –2
	decline	–2 / 6	0 / 0

"Very good, young man," praises Tucker. "You've recognized that the two games are identical as far as their structure is concerned. From a mathematical point of view, it is the same game. Both tables depict the same dilemma: if Al and Capone in the prisoners' dilemma or Ann and Bob in the 'investment dilemma' were to cooperate, they would gain the greatest joint benefit. They won't do this, however, because they are heedlessly intent on their own gain, which would be greater than their share of the joint gain. By doing so, however, they harm themselves and the other, too. Nevertheless, there isn't a sensible reason for either of them to depart from a strategy that actually harms both of them."

Albert Tucker's presentation of the prisoners' dilemma resonated far beyond the lecture hall at Stanford. In tens of thousands of negotiations since, variations of this game have been invented and considerations dreamt up in order to blunt the dilemma's edge. As the student Ann rightly said: when we view it from outside, we shouldn't even talk of a dilemma but rather of the rational behavior of the players. But how can their mutual infliction of harm be prevented? A question that since then has occupied those whose aim is not to win games but to analyze them.

The obvious answer is that not allowing communication prevents potential cooperation between the players. That is true, but it is only part of the issue.

Let's imagine that, during the night and before the district attorney enters his cell again the next day, Capone receives a secret message. On the dirty paper, in almost illegible writing, it says: "I won't squeal. Al." What will Capone do? Should he rely on Al, whom he has known forever as a devious rogue, to keep his promise this time? Can he even be sure that the note is really from Al, and isn't just a nasty trick played by the prison warden, who has never been able to stand Capone? When the district attorney questions Al, Capone won't hear what they are saying. He will perhaps wrestle even more with the dilemma than if he hadn't received the secret note. Besides, can Al, for his part, even be sure that the note will find its way into Capone's hands? Won't Al ultimately choose to spill the beans rather than keep quiet?

Even if, before they are individually questioned, they could meet and guarantee to each other to keep their mouths shut, there is a good chance that, during the questioning itself, with Capone all alone with the district attorney, he will confess. His gain when confessing is simply greater, regardless of how Al behaves.

For Ann, who believes in the enforcement of justice, an unsatisfactory situation would arise if the two crooks, while in possession of illegal firearms, had merely planned to rob the bank, but other robbers had actually carried out the crime. In this case, the district attorney would expose to the dilemma two people who hadn't actually committed the crime he suspects them of. But regardless of whether they robbed the bank or not, the game table that illustrates the prisoners' dilemma requires that they should confess, in which case not only justice but also truth would be a victim of the prisoners' dilemma.

Moreover, it is rather disconcerting to learn that the prisoners' dilemma is not limited to criminals. Its variation in the form of the investment game shows that, in theory, this dilemma is involved in every business transaction. Naturally, the game is completely unrealistic in this particular form—nobody would ever be able to invest nothing and still have a share of the profits. But we can assume that every sensible person who is active in business would aim to minimize their investment and maximize their return and at times this aim does violate decent moral standards. At least the prisoners' dilemma does explain how such intentions that are detrimental to joint business necessarily come about.

But how can we escape the dilemma?

At this point, time becomes our focus once more. We already know that, in order to remain on the "winning" side in the "game" of business, it is always imperative to choose the right moment. How we can apply this realization to the investment game and thereby attempt to bypass the dilemma inherent in it is another fascinating story.

12
PLAYING WITH PROFIT

BERKELEY, NEAR SAN FRANCISCO, CALIFORNIA, 1980

"Tit for tat."

In common parlance, this is what people say for the old Latin axiom of law "quid pro quo," by which a person who gives something is supposed to receive something proportionate in return. "Quid pro quo" sounds a bit better than the other Latin maxim "Manus manum lavat" (One hand washes the other), since the latter contains the suggestion that dubious machinations are involved, not worthy of honest, upstanding business people. The English saying actually comes from "tip for tap," meaning "blow for blow," and it certainly has a more martial sound than the Latin "quid pro quo."

No matter how simple this maxim may sound, it turns out to be extremely effective when one is attempting to take the edge off the prisoners' dilemma.

As Albert Tucker leaves the lecture hall after his talk presenting the prisoners' dilemma, the young student Ann pushes her way through the crowd to him and asks if she can have a word.

"In answer to an objection by one of my fellow students, you said that you chose to ignore the possibility of Capone's friends taking revenge on Al, should he blow the whistle on Capone. I understood that with the prisoners. In any case, having confessed, Al could request police protection after his release. So in this matter, I can follow your thinking in every respect. In the investment game, however, the situation seems to be more complicated."

"Why?" asks Tucker with interest.

"Because the two business associates involved in the game don't make this decision about a potential investment just once in their lives; rather, they make decisions on a regular basis, sometimes even daily, that may not be exactly like the one in your investment game, but are roughly the same. If people know that one of the business partners constantly declines the investment, but gains the greatest possible benefit from it for himself, then they will soon exclude him from the game. You see what I mean, Professor Tucker—with the investment game, it is surely necessary to take into account how a player's selfish decision, which benefits him more in the short term than one aimed at the common gain, will affect him in the long term."

This excellent consideration prompted the game theorists to create a new version of the investment game, where the two players play the game not just once, but more often.

"How often?" That was the first question asked.

If it is agreed with the players that the investment game should be played ten times, it necessarily follows that both players will decline the investment in the tenth and last round, since nothing comes after this round and declining the investment brings the individual player—regardless what the other one does—a greater gain. While this is zero, since the other player will also decline, it is at least not negative, which would be the case if the other player declines the investment, but one makes the investment oneself.

If the players know, however, that they will gain no profit from the tenth and last round, then they don't even need to play it, since it is purely a waste of time. From their point of view, therefore, the ninth round is actually the last one. So the players' decisions in the ninth round will be made as though it is the last round—both of them will decline the investment and gain no profit from the game. Like the tenth round, the ninth one will have become superfluous.

In this way, the intention to decline the investment eats its way back from the tenth round to the ninth, to the eighth, seventh, and on to the first round, and through all ten rounds, it proves to be the sensible course of action. And the dilemma is that nothing happens in any of the ten rounds. Never is any profit paid out.

Viewed from a mathematical perspective, one will experience the same disappointing lack of action if one arranges with the players to repeat the investment game a hundred times. In real life, one will almost certainly find nobody who would go along with such a pointless and time-consuming undertaking. The question "How often?" must therefore be answered differently.

The players must not know how often the investment game is to be repeated. The game master can decide on a whim how many rounds are to be played, for example. A neater and more certain solution—after all, one never knows if one of the players has bribed the game master to get the number of games out of him—is to throw a dice after each round. Only if it is a six is the completed round the last one—otherwise, another one is then played. If the players are patient enough, it can be agreed to spin a roulette wheel instead. Only if the ball lands in the slot marked zero is the completed round the last one—otherwise, another round is played.

Before making abstract observations with pen and paper, the game theorists experimented with real people and real money—although only with a few dollars, rather than tens of thousands. It did indeed turn out to be the case that, when playing several rounds of the investment game without anybody knowing which round was the last, the players would take great risks when investing. They would invest rather a lot of money, even when they were sitting in separate rooms, like the robbers Al and Capone, and couldn't communicate with one another or even see each other's faces. Some of the players managed to snag considerable sums. Was it pure luck, a fortunate coincidence, or had they devised a cunning system for the game? Is there in the iterated investment game—as we call it when the game is repeated with an unknown number of rounds—a strategy that, unlike the pointless systems in roulette, is actually successful?

Over the course of subsequent investigations, experiments with real people and real money were dispensed with, since the undertaking would have become rather expensive. In 1978, the thirty-five-year-old political scientist Robert Axelrod, from the RAND Corporation, suggested using a computer to come to grips with the problem. He wrote to game theorists, inviting them to take part in a competition, for which each of them had to submit a program that would simulate a player's behavior in the iterated investment game. Such a program can be designed to be very simple. It can, for example, be made to ensure that the player it simulates constantly makes the investment. If two such programs play with each other, this naturally means a huge profit for both—but only in this particular case. Or the program can be designed to ensure that the player it simulates constantly declines the investment. This means that one is at least on the safe side and will never lose, and if one is lucky and the opposing program risks an investment in some of the rounds, one can even win some money. Alternatively, the program can have a random generator built in, as though the simulated player throws a die before each round and the number he throws determines whether he makes the investment or not. If, for example, the investment is only made if a six is thrown, there will be a probability of approximately 16.7 percent that the program will invest, and it will otherwise decline. But the program can also undertake to invest in the next round if an even number is thrown and to decline the investment if it is an odd number. In that case, it would invest randomly, but with a fixed probability of 50 percent.

Of particular interest are those programs that make a decision for the next round based on the previous decisions it, or the other player, has made. Because the present is here determined based on the known past and in the hope of a better future, time first becomes a factor. We can recall Benjamin Franklin's phrase, "Time is money." It is time that allows us to adjust our future decisions according to the other player's previous behavior.

At any rate, fifteen different programs were submitted to Axelrod, most of them of a rather complicated design. Programs independent of time, one that always invests, one that always declines, and those that invest or decline at the direction of a random generator—all of these were included in the list. Axelrod then made each individual program play against itself and against all the other programs. The program that achieved the greatest overall profit in all of the iterated investment games would, he thought, reveal the strategy most worthy of adopting when playing the investment game.

One of the fifteen submitted programs did indeed emerge as a clear, undoubted winner. It was created by Anatol Rapoport, a versatile American academic who had been born in tsarist Russia in 1911. The most notable thing about the Rapoport program was that it was by far the most simple of all those submitted. It makes the investment in the first round and then sticks solely to the rule of quid pro quo. In other words, if the other player has made the investment in the previous round, the Rapoport program player does so in the next one. If the other player has declined the investment in the previous round, the Rapoport program player does likewise in the next one. Tit for tat, exactly as the saying goes. Risking an investment in the first round and then in every round doing what the other player has done in the previous round could be expressed in Latin with the saying "Primum pro, tunc quid pro quo," or in English with "First be nice, then tit for tat will suffice."

In order to understand the attraction of the Rapoport program, we can start off with the investment game table from the previous chapter, forgetting the tens of thousands written in the table and replacing dollars with more valuable gold ducats. If both players invest in a given round, each of them gets four ducats. If both players decline the investment in a given round, both of them get nothing. And if one player invests, but the other declines, then the first player will have to accept a loss of two ducats, while the other one receives six ducats.

Now let us consider various scenarios with the Rapoport program. We will call the player who sticks rigidly to the Rapoport program

Anatol, since that is Rapoport's first name. And we might as well call Anatol's playing partner Max, since that is the name of the hero's friend in Arthur Schnitzler's play *Anatol*.

If Max invests in every round, Anatol will also do so. In each round, both players will each win four ducats, and it will be in their interests to play as many rounds as possible.

If, on the other hand, Max declines the investment in every round, Anatol will do likewise, apart from in the first round, when he loses two ducats. In this case, it doesn't make any difference to the players how many rounds there are in the investment game, because nothing will happen after the first round. Max takes home a profit of six ducats, hardly a thrilling result if they have played a large number of rounds, while Anatol is left with an acceptable loss of two ducats—acceptable, because his loss could have been much greater over the course of all the rounds.

Things get interesting if Max's decision whether to invest in the next round or not is completely unpredictable. Let us assume that Anatol and Max have the same number of ducats in their wallets at the beginning of the game: if, in the first round, both Anatol (following the pattern of the Rapoport program) and Max make the investment, they each receive four ducats and their wallets are still equally full. If, on the other hand, Max declines the investment in the first round, there is then a difference of eight ducats in how full their respective wallets are, to the detriment of Anatol. This difference will remain until Max finally decides to make the investment in the coming round. In keeping with the formula of the Rapoport program, Anatol declines the investment in this coming round. Max is thus penalized and the contents of their wallets are now equal once again after this round, and the following rounds continue as though they were just beginning the game. It would only be bad for Anatol if the game were to be interrupted while he is eight ducats behind Max, but the difference can never be greater than this.

The ideal scenario is when Max also plays according to the quid-pro-quo rule—in other words, the Rapoport program is allowed to play with itself. In this case, it is exactly the same as with two players who always make the investment—they both earn from the iterated investment game a guaranteed and significant profit.

We tend to empathize with Anatol, who plays in accordance with the Rapoport program, for four reasons: firstly, he is an optimist—when he begins the game, he acts on the assumption that the other player will make the investment like he does. Secondly, he is uncomplicated—his behavior is, in the best possible way, simple. Thirdly, he doesn't allow himself to be taken advantage of—if his playing partner sends him down a dead end, he pays him back for this in the next round. And fourthly, he doesn't bear a grudge—as soon as the other player decides to collaborate, Anatol is ready to cooperate too.

On top of this, Anatol does not set out to gain more profit from the iterated investment game than Max. He could never achieve this while sticking rigidly to the Rapoport program. It is therefore significant that this game is not a zero-sum game. In zero-sum games with two players, one person must necessarily be the loser, whereas in the iterated investment game with two players, the second one can win practically the same amount as the first. Anatol is thus not characterized by excessive ambition, and his goals are modest. It is enough for him to be a winner along with the other player. And his modesty is wise, since he can therefore never lose too much when playing against crooks.

Having said that, this does not necessarily mean that Rapoport's program offers the perfect solution in every respect. Naturally, the player who always declines the investment wins the most, if playing against a good-natured but simple soul who, regardless of all his losses, makes the investment in every round. But such a scenario practically never occurs in real life.

There is a more realistic danger that, when Anatol is playing with Max, the latter, inadvertently or because of a misunderstanding rather than out of spite, suddenly declines the investment. Immediately after this, Max determines to dutifully follow the quid-pro-quo rule like Anatol. But the harm has already been done. Anatol doesn't invest in the next round—since Max had mistakenly declined in the previous round—while Max then makes the investment in the following round, because he has decided to adopt the Rapoport program. Because Anatol then declines, in line with the program's strategy, this then forces Max also to decline in the round after that, when Anatol, on the other hand, makes the investment again. And thus this vicious circle of alternate investing and declining continues until the end of the game, and both players will no longer be able to gain the profit that would otherwise have been possible.

The London-born biologist John Maynard Smith suggested as an alternative the program "Tit for Two Tats": only if Max declines the investment two times in a row will his fellow player John, who is following the "Tit for Two Tats" program, decline the investment by way of punishment in the next round. This means that a "mistake" by Max is no reason to end up in the vicious circle. On the other hand, if Max is crafty and realizes that John is acting in accordance with the "Tit for Two Tats" strategy, he could fleece him by declining the investment time and again, but never twice in a row. The "Tit for Two Tats" program is therefore not as failsafe against crooks as the Rapoport program.

It may thus be wiser for Anatol to judge, based on the rounds played so far and Max's playing behavior in them, how likely it is that, when Max declines, this is a deliberate ploy or perhaps just a forgivable misunderstanding after all. Anatol therefore modifies the Rapoport program to "Sometimes Tit for Tat": only with a certain probability will he decline the investment in the next round if Max has done so in the round before. He will be wise enough to set this probability at 100 percent to begin with—after all, we never know what kind of person we are dealing with at the outset. But if, over the course of many rounds, Max proves to be a particularly eager investor, the probability of unforgivingly hitting back in kind after he suddenly declines will become significantly less than 100%. If these aberrations on Max's part mount up, however, the probability will very quickly rise toward 100 percent again. This error-tolerant modification of the Rapoport program takes the time factor into account even more than the program itself.

The deliberations suggested so far show how a vast number of variants of the iterated investment game, complete with related strategy programs, were brought into play by those researching the matter. It is no coincidence, therefore, that the aforementioned John Maynard Smith was a biologist: the iterated investment game cannot only be carried out with just two players—many more can also be involved. Suddenly, populations form, following strategies they find successful, exactly as can be observed in the biological evolution of species in the history of life on Earth, something demonstrated by Karl Sigmund with the help of numerous fascinating details in his marvelous book *Games of Life.*

It is equally remarkable how human nature is reflected in the strategy programs for the iterated investment game. There are those who are distrustful, who do adhere to the quid-pro-quo rule, but begin the first round by declining the investment; those who are unforgiving, who make the investment to begin with, but then, as soon as the other player declines, decline in all further rounds until the end of the game; those who are vengeful, who will not invest again until their opponent's profit, gained by declining the investment, has been wiped out; those who are conformist, who play in the next round like the other player has done for the majority of the previous rounds. And, as already mentioned, there are those whose playing behavior is completely independent of time, those who are unpredictable and let chance decide whether they invest in the next round or not, those who are stubbornly malicious and permanently decline the investment, and those who are incorrigibly naive and make the investment in every round, regardless how much they have been deceived before. All of these and many more types of playing partners face wise Anatol, who sticks to the rule of quid pro quo.

It was no surprise that Anatol Rapoport should be the one who invented this highly successful program. Born in the Ukrainian city of Losovaya, he was eleven when his family emigrated to the United

States. His father, who had a major influence on his son, was one of the Russian intellectuals typical for the period, who were fascinated by the West's Enlightenment and pursued the dreams of a world improved by reason that were propagated by several of their number. Rapoport's father recognized his son's musical talents and encouraged him to train as a pianist. The second person who left his mark on Rapoport's development was his piano teacher Glenn Dillard Gunn, who taught at a small Chicago music school and of whom Rapoport said,

> He was in no way an outstanding musician, certainly not an outstanding pianist. But he was a youthfully enthusiastic lover of music at age fifty-three, an apostle of late romanticism (Berlioz, Liszt, Wagner). He was solidly American, smoked cigars, lived in a suburb, was possibly (though I am not sure) a member of a lodge. But he had studied in Leipzig, spoke fluent German, and engaged me in prolonged conversations about the world of music and musicians during the European *fin-de-siècle*.[1]

In 1929, Rapoport traveled to the center of music, Vienna, and studied piano and composition there at the State Academy of Music and Performing Arts until 1934. Alongside his studies, he worked as a correspondent for the American magazine *Musical Courier* and appeared as a concert pianist and lecturer in Europe and the United States. In 1934, he switched to mathematics and studied the subject in Chicago, getting to know the third key role model of his life, his doctoral supervisor Nicolas Rashevsky:

> Like Gunn, he had a precious sense of humour and a prodigious memory. He could quote pages of Russian classics and recite biting satirical poetry by Russian students.
>
> In America Rashevsky went off to work for Westinghouse Electric. When the Depression hit, they fired him. He found a niche at the University of Chicago. There he continued to develop mathematical biophysics, a new field he had founded while still in Pittsburgh.[2]

Writing about his relationship with Rashevsky, Rapoport says, "I worked with him for seven years in Chicago. Although Gunn was thirty-seven years older than I and Rashevsky only twelve, my relations with both men were quite similar—a mixture of teacher-student and peer-friend relation, which meant much to me and, I think, also to my mentors."[3]

It was as a result of his involvement with biophysics that Rapoport developed the interest in analyzing conflict and confrontation that would influence the rest of his long life. In his way of thinking, games are the golden mean between debate on the one hand, where one elucidates the merits of one's own position to the other person, and combat on the other, where the other person becomes an opponent who must be forced into submission. Games are a trial of strength that follows fixed rules; they do not end with one of the players losing face, but with the simple diagnosis of who has won or lost, and how much.

Rapoport was active for many years in various peace movements. He was familiar with the "game" of the Cold War, which was not dissimilar to the prisoners' dilemma: if the United States and the Soviet Union did not stop their military buildup, their mutual threats could become a full-blown catastrophe at the slightest provocation. The optimal solution would have been the mutual disarmament of the great military blocs, but the dilemma lay in the fact that one-sided disarmament would have given the opponent an unwanted degree of power.

The good news is that history is endless. The "game" of military politics is not only played once, but is repeated time and time again. It is therefore possible to circumvent the dilemma with sensible strategies, with the quid-pro-quo rule brought into play by Rapoport being one of these.

The bad news is that history is endless. New playing partners constantly appear to replace those that disappear. New stakes are constantly required, and they are at times so high that even a single loss in one of the rounds of the "game" can signify the destruction of the player. Reality is much more varied than what could ever be modeled on a game board.

Rapoport was appointed professor of mathematical biology at the University of Michigan and then, from 1970, professor of psychology and mathematics at the University of Toronto, where he also became professor of peace and conflict studies. His wide-ranging research activities took him all over the world, including to Vienna, the city where he had once studied music, and he spent three years there as the director of the Institute for Advanced Studies.

And so we return to Vienna once again, the city that we lost sight of in our story with Menger, Gödel, and so many others after the end of the 1930s.

13

PLAYING WITH THE POLICE

VIENNA, 2002

"There's no such thing as a free lunch."

The discussant ends his contribution to the forum with these words and looks benignly at the evening's speaker. "Nothing is free—even the coffee costs something here," adds another, and a third person provides a Russian version of the saying: "Free cheese is only found in a mousetrap."

"I certainly agree with you," says the speaker, and thus draws to a close one of the most fascinating events in the "Café Philosophique" series, held regularly between 1997 and 2007 at the Café Prückel, one of the last great coffeehouses remaining on Vienna's Ringstrasse, the magnificent boulevard that circles the inner city center.

During this period, Karl Riesenhuber and Reinhard Hosch, both Viennese intellectuals and teachers of Romance languages, organized the Café Philosophique on a monthly basis: anybody who was interested in the topic advertised was invited to come to the Café Prückel, sit at one of the tables moved together for the event, order a coffee or something more substantial, and wait until the event began at five o'clock and the speaker chosen to provide the impulse for the philosophical discussion was introduced. Riesenhuber and Hosch understood "philosophy" not as the subject taught at the universities but in a much wider sense: as a kind of enlightened reasoning without barriers. This is how it was viewed by Marc Sautet, who founded the first Café Philosophique at the Café des Phares on Place de la Bastille in Paris, although only he himself served as the speaker who stimulated the discussion. Those taking part in the Café Philosophique were dilettantes of thinking, with "dilettante" meant not in a derogatory sense of an incompetent person but in a literal sense to mean a true lover of thinking.

On this occasion, in June 2002, the speaker, though more of an impulse-provider, is Karl Sigmund, his subject being "The Arithmetic of Morality." Sigmund, too, is not a philosopher by profession—he is professor of mathematics at the University of Vienna. He is one of those who, in the second generation, so to speak, after the Third Reich's devastation of cultural, scientific, and intellectual life, helped to restore the outstanding reputation of mathematics in Austria.

The state of the university, and that of the mathematical institute, was indeed highly inauspicious shortly after the end of the Second World War. One might have expected that Austria, once reconstructed, would make every effort to invite the country's emigrant intellectual giants to return home. Karl Menger, for example, in faraway Chicago, where he was attempting without much success to organize a "Chicago Circle," would have liked nothing better than to be summoned back to Vienna.

"Menger wants to become professor of mathematics in Vienna again," whispers an undersecretary to his department head.

"Menger? Menger?" The latter tries to recall the name, before finally remembering: "Ah yes, his father was a most peculiar Liberal and his mother was a Jew. That's why he had to leave the country. And just now, when we can't assign any posts at all, he wants to return? That doesn't suit us at all."

"Think of our reputation, sir!"

"But it will be too expensive. And I don't think he knows what he would be letting himself in for. With Vienna bombed to pieces, he wants to leave the wealth of America behind now? He can't be serious."

"Besides, I found out that he voluntarily gave up the professorship in 1938," says the undersecretary officiously.

"Well done," says his superior, visibly pleased. "Then we can write to him and say something along the lines of: 'We regret that, on account of your having previously unilaterally renounced the professorship, formal grounds render it impossible for us to follow up your esteemed suggestion.' Please be good enough to compose an appropriate reply straightaway."

In scenes such as this, provincial small-mindedness proved to be shamefully insulting, not only in relation to Karl Menger's desired return to Austria but also to that of many other notable figures. Not one of the decision-makers of the time recognized the irretrievable opportunities that they let slip away with such affronts.

Nevertheless, there was a quartet of professors in the field of mathematics at the University of Vienna who, on account of their blameless behavior in the time before, their remarkable qualities in research and teaching, and their dedication, managed successfully to rebuild the subject's reputation in Vienna.

The first was Nikolaus Hofreiter, a former assistant of Furtwängler's who had carried out research in Furtwängler's specialist field and now focused his full attention on teaching and organizing the faculty: an exceedingly careful, assiduous, conscientious, almost fastidious man—the very definition of an Austrian public official, though in the best sense.

Then there was Johann Radon, the oldest of the four: he had studied in Vienna under Hans Hahn, among others, written his doctoral thesis under the supervision of Wilhelm Wirtinger, and, as a student in Göttingen in 1910, attended the lectures of David Hilbert, the high priest of mathematics at the time. Via Hamburg, Greifswald, and Breslau (now Wrocław), he came to Vienna directly after the war. His mathematical research made him world famous in specialist circles, but his discovery of the theoretical foundations of computer tomography long before the practical implementation of this invention is perhaps particularly worthy of note. He was a kind and likeable man, equally popular with his pupils and colleagues—a noble-minded character.

The third man was Edmund Hlawka, a highly talented and versatile mathematician for whom Hofreiter, his colleague of almost the same age, served as his dissertation supervisor, but who in actual fact was perfectly capable of developing his highly original ideas by himself. When asked who his actual teacher was, he would refer to Carl Ludwig Siegel, who taught at the Institute for Advanced Study and in Göttingen and was probably the greatest figure in mathematics in the second half of the twentieth century. The University of Vienna was extremely fortunate that Siegel did not succeed in tempting his favorite student Hlawka away from his home city.

The fourth member of the quartet was Hlawka's friend Leopold Schmetterer, who had studied with him and also done his doctorate under Hofreiter, before then serving as Radon's assistant. He became professor of mathematical statistics in Hamburg, where Emil Artin, who had returned from Princeton, was a colleague of his. After Radon's death, he took over his chair in Vienna and endeavored to place statistics, a subject that had been only very superficially taught in Austria since the end of the war, on a solid mathematical footing. One of his most outstanding students was the very Karl Sigmund who is now, in 2002, speaking to an interested but non-specialist audience about the "Arithmetic of Morality" at the Café Philosophique at the Café Prückel.

Sigmund is better qualified for this than practically anybody else in Vienna. As an acknowledged expert on probability theory, he has turned his attention to game theory and, together with the biochemist Peter Schuster and following the work of John Maynard Smith, has gained impressive insights into the evolution of life, using mathematical means. Among his best students are Josef Hofbauer, the coauthor of his book *Evolutionary Games and Population Dynamics*, and Martin Nowak, who taught with great success at Oxford and Princeton and is currently director of Harvard University's Program for Evolutionary Dynamics, a research and teaching institution established in 2003 and dedicated to the study of the fundamental mathematical principles that guide evolution.

"It is not morality that I wish to define," says Sigmund, beginning his introductory talk at the Café, "but rather how we believe we can recognize moral behavior using simple games. And I'm afraid I must also disappoint you as far as the games are concerned. You will not hear the winning formula for games like chess, Go, tarock, or poker. There isn't such a thing with most games. They are simply too complicated. I am merely going to tell you, using very simple games that may initially seem rather artificial, how the structure of games can be explained with the help of mathematics."

Naturally, the prisoners' dilemma and its variant, the investment game, form the core of his short speech. During the lively discussion that follows, Sigmund also analyzes the iterated investment game and the quid-pro-quo rule in a very similar way to how these were presented in the previous chapter. In light of the crowd of some seventy interested listeners in the coffeehouse, he decides to recount the following form of the investment game:

"Imagine that we are playing here a large number of rounds of the investment game. In each round, I come to each and every one of you with a box, and you can either put ten euros inside or decline to make this offering. Afterward, I increase the money collected by 50 percent and share out the total amount among all of you equally. I keep none of it for myself. Which of you would entrust the ten euros to me?"

Everybody present raises his or her hand, with the sole exception of one man, who is sitting at the table next to Sigmund. "Very good, thank you," says Sigmund encouragingly. He then points to the man next to him: "This fellow player of yours did not put up his hand and thus made a clever decision. All of you can expect a small profit—if all of you put up your hands, the net profit is five euros and if, like now, the vast majority of you put up your hands, it is a little under five euros. For this man next to me, however, who did not put up his hand, a larger profit than yours is in store—almost fifteen euros, since so many hands were raised."

"But that shows a lack of solidarity," complains somebody at a table in the background.

"Quite right, and there will be a price to be paid in the subsequent rounds of the game," continues Sigmund. "It has been tried out time and time again: after each subsequent round, more and more players decide not to put any money into the box. Initially, this is because they recognize that they thereby receive more than the difference between the deposit of ten euros and the amount they get back afterward. However, as more and more players decide not to contribute to the pot, this difference becomes correspondingly smaller, much to the annoyance of everybody. Finally, those who donate ten euros actually make a loss and, from this point on, practically everybody declines to put ten euros in the box."

"The reason why people selfishly decline the investment certainly has to do with the fact that nobody knows whether the others have donated the money or not," suggests one lady.

"Yes and no," answers Sigmund. "If one has to show everybody present whether one invests the ten euros or declines this investment, then the moral standard—if I may put it that way—that is to say our keenness to donate the money, remains high. But as soon as the others discover that somebody has no qualms about saying no when the game master asks for the ten euros and that person receives a larger payout than those who were happy to donate, then the same effect will quickly occur. In the end, nobody will give anything. However, another addition to this game suddenly alters everything—if the game master, after collecting the money, allows players to come forward who are interested in punishing those who have declined to give. One is allowed to play 'Police' at the same time. However, there is a catch: a player who acts as the police and demands from one who has declined to give that he hands over a fine of twenty euros to the game master must himself pay five euros, which he won't see again. It is unbelievable how the introduction of this ability to punish others changes the course of the game. The players are delighted when punishment is thus permitted. They happily pay five euros for the privilege of being able to point at somebody who has declined to give ten euros and shout out: 'He didn't invest anything, he has damaged our prospects, and because we receive less of a profit, he needs to be punished!'"

Karl Sigmund holds up his right hand and then, as he allows it to fall to the side, he says, "This is how the players' 'moral standard' crumbles when there are no 'police.' It ends up at rock bottom and stays there forever. But if players are allowed to play at being policemen and punish the offenders for five euros each, the 'moral standard' falls to begin with"—Sigmund lets his right hand fall again, but lifts his left hand a little up into the air—"but the number of those who want at all costs to hand out punishment rises correspondingly. If those declining to give realize that they can be punished, they begin again to dutifully invest"—Sigmund's right hand rises again—"and the number of 'policemen' naturally decreases." His left hand falls again. "As soon as the shrewder players notice that there are no more 'police,' they take the risk of not investing, because they thereby hope for a greater profit, and so the whole to-and-fro between the decline in 'morality,' the increase in 'policemen,' the restoration of 'morality,' and the decrease in 'policemen' is repeated over and over again."

Reinhard Hosch, one of the organizers of the Café Philosophique, has a question: "At school, it is drummed into teachers that they shouldn't punish the children who don't complete their tasks, but much rather should praise those that do. 'Praise, not punishment' is the motto of benevolent educational theory. One could try that in this game, too: instead of the 'police,' we could introduce a kind of 'praise committee.' Players would come forward after the money has been collected and would point to somebody who has dutifully invested the ten euros and say: 'He invested, he benefited the common interest—because we get a greater profit thanks to him, he needs to be rewarded with, let's say, five euros!' Has that ever been tried?"

"It has indeed," answers Karl Sigmund, "and the sad truth is that it didn't help in the slightest. With praise alone and without the introduction of the police, 'morality' sinks inexorably to the bottom."

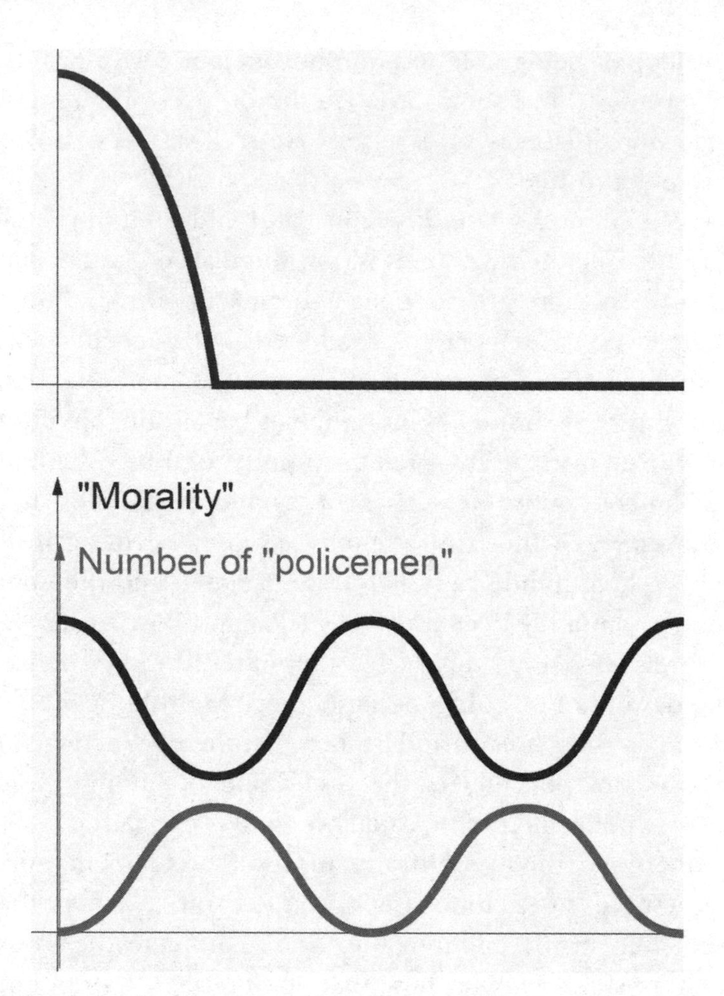

Figure 13.1: The top sketch shows how swiftly the "moral standard" declines in the investment game with many players; to be more precise, it shows that the number of players willing to invest quickly falls to zero and remains there thenceforth. The bottom sketch shows that the number of players willing to invest oscillates depending on the number of "policemen" who act to punish those who decline to invest.

"I still have two questions about your game," a young lady comments. "The first one concerns dividing the profit among the players equally, regardless of whether they have invested or not. Why is there such a rule, which strikes me as being unfair? And the second question concerns the 50 percent increase to the invested money: where does the game master get this extra money?"

"Interesting points," admits Karl Sigmund. "Let me begin with your first question: Each of us, as soon as we take on employment, is involved in such a sharing of profit. Take me as an example: on my monthly pay slip, I can see that the work I have done—in the words of the game, my investment—is valued at four thousand euros. That is my gross monthly salary. Further down on the slip, however, I see that only a net amount of two thousand euros remains. Where has the other two thousand euros of my four thousand–euro investment gone, I ask in amazement. And I hear that, whether I want to or not, I have given this amount to the state in the form of taxes and fees.

"A lot of this money is given by the state to those who do not currently contribute to an increase in the gross national product, who do not—to stick to the imagery of our game—make an investment, as I do with my work. 'Currently' is the key word, here, since these other people have already invested, or will invest, at some point. We're talking about children and young people going to school or university, who will increase the gross national product in the future. Or about people who are dependent on government support, who have worked in the past and will hopefully return to employment again soon. Or those who are retired, who have already contributed a great deal to society. It is for all of these people that we pay taxes and fees. In addition, we pay for investments for the future, which may perhaps only come to fruition when I myself am no longer in active employment.

"Here in Austria, there is a consensus that it is the state's duty to provide such social benefits and centrally managed investments. In the United States, on the other hand, more emphasis is placed on individual initiative and there is less state redistribution. As a result, a greater proportion of each person's gross salary remains as net income. In principle, however, a properly functioning national economy is founded on the premise that the state is the game master who defines the rules and the workers, whether self-employed or not, and all other members of society are the players who make an investment with their work and contribution, just like in our simple game where you have the chance to invest ten euros—and will generally do so if you can.

"Obviously, reality is much more complex and diverse than our simple investment game. But one essential aspect is made clear: that in order to increase the quality of life for all members of society, it is important that individuals do not keep the entire profit they have helped create, but share it with others.

"And now to answer your second question: The game master is able to increase the amount of money because he draws it from the future. It is economic growth that causes an increase in value and thereby of money. That is why it is so important that we live in times of long-term economic growth, at most only interrupted by short periods of recession. Our great technological achievements indicate that this could continue, and that is our great hope. But unfortunately it is by no means certain. If the capital used for investments does not yield any interest in the long term, we will enter an era where the economy is dominated by zero-sum games, with all the terrible consequences that result from that, because a player can only win in a zero-sum game if he harms his competitors. There can be no win-win situation. If you come second in a zero-sum game, you have already lost."

As an experienced speaker, Sigmund knows that it would be unwise to close the discussion at this point. He therefore decides to finish off by talking about a completely different game.

"I must introduce you to another game, called 'Ultimatum.' It is one of the most banal and, at the same time, bizarre games imaginable: two players—they don't need to know each other and can even remain in separate rooms—and a game master are involved.

"First, a coin is tossed to decide which of the two players has the first move. The game master then puts a hundred one-euro coins on the table in front of the player who is first and offers him the chance to take some of them and leave the rest for the other player. The snag is that the other player knows that a hundred one-euro coins are to be shared between him and the first player. He can accept the remaining money left by the first player, in which case the money has thus been shared between them and the game is over, or he can reject the proposal, in which case neither player receives anything, the money remains with the game master and the game is also over. Allow me to demonstrate this game with the two gentlemen who invited me here. Mr. Riesenhuber, you have the first option and must decide how many of the hundred one-euro coins you want to take for yourself."

"I'll take exactly half, so fifty euros."

"A fair-minded man," says Sigmund approvingly, and asks Reinhard Hosch if he is happy to accept the other half. Hosch nods his assent and, before he can speak, Karl Sigmund continues with his explanation:

"It goes without saying that the second player agrees to such a proposal. Nobody would be able to understand it if he didn't. Over many thousands of rounds in which the Ultimatum game has been tried out, in every possible corner of the world and in the most varied cultural circles, it has been demonstrated, regardless of differences in gender, religion, education, or age, that the second player accepts the share left by the first player if what is left is not too far off a fifty-fifty share. He is generally prepared to accept forty euros, for instance. But an unfair proposed share—ninety euros for the first player and ten for the second—is rejected out of hand by the second player, indeed viewed as an indignity.

"From a purely mathematical point of view, such behavior on account of a division that is considered unfair is astonishing. A player who is only geared toward profit would think that the first player should really take ninety-nine euros for himself and leave one euro for the other. After all, one euro is more than nothing, thinks the 'homo oeconomicus,' the player geared solely toward profit, and so he believes the other player will accept such a small profit rather than go empty-handed. But far from it! Man is not a 'homo oeconomicus.'"

"Not even when mathematicians are playing?" asks a wit in the audience.

"Not even with those fellows," answers Sigmund. "On the other hand, when this game was conducted with a particular tribe of South American natives, a highly unusual kind of playing behavior was witnessed among the people there. Regardless of what the first player's proposed share was, even the completely fair-minded fifty-fifty division, the second player always rejected the proposal. At first, nobody could understand why the players in this tribe behaved like this. It was only upon enquiry that the answer to this mystery was found out: the game played there was managed by a young woman, and the natives believed that this woman, out of the goodness of her heart, had donated the money that she put on the table from her own pocket. And none of the players wanted to take this money away from her."

The man at Karl Sigmund's table who had previously not raised his hand when the audience was asked who would be prepared to invest ten euros now speaks up:

"It is indeed fascinating to learn that people do make decisions for reasons other than the pursuit of even more money. But what those reasons are and how strongly grounded they are is still a mystery. On top of this, I think I may have found the Achilles' heel of the Ultimatum game—it has absolutely nothing to do with the life that we have to or want to lead. It is a completely unrealistic game. Because you never ever get offered money for nothing, not even one paltry euro. All money must be earned in some way. That's one of the fundamental laws of economics: there's no such thing as a free lunch."

14

PLAYING WITH INFORMATION

NEW YORK CITY, 1990

"Switch, and your chances of winning are doubled!"

When Marilyn vos Savant wrote this sentence in answer to a question in the American Sunday magazine *Parade*, she had no idea what an outcry it would cause. The magazine is enclosed with more than 640 newspapers each Sunday and so has a huge readership. One of the most popular columns in *Parade* is called Ask Marilyn. The author, Marilyn vos Savant, is known to have the highest IQ ever measured and answers readers' questions of every imaginable kind, with these questions and her answers forming the Ask Marilyn column.

She knows, for example, why there seems to be a solemn stillness outdoors when snowflakes are falling thickly from the sky on winter nights; why eggs become hard when they are boiled, although materials usually turn from a solid to a liquid when heated; or why rockets are propelled by pushing out gases in empty space, even though there is nothing there against which to push. These and similar questions, sent in by readers from all fifty American states, are found along with her answers in the Ask Marilyn column.

It is the edition of September 9, 1990, in which Marilyn vos Savant gives a surprising answer to a question posed by Craig F. Whitaker from Columbia, Maryland.

In order to gain a better understanding of Whitaker's question, we must first say something about the TV show *Let's Make a Deal.* For literally decades, this show has been broadcast on the major national American TV channels, featuring a skillful host who conducts games with candidates—in its heyday, the show's host was the Canadian-born Monty Hall, whose real name was Monte Halparin. The games overseen by Monty Hall with people chosen from the audience are full of variety, but always such that there is no shame in losing. There are no questions demanding intelligence or skill, and it is always a matter of pure chance whether luck smiles on the candidate or not.

The most popular and oft-repeated of all the games on *Let's Make a Deal* is known as the "three-door game." The mere announcement of this game makes the audience cheer with joyful excitement. Monty Hall and Lucy, the candidate, are standing in front of three closed doors.

"Lucy, here are three doors," begins Monty Hall. "Behind one of them—either the one with the number 1 or the one with the number 2 or the one with the number 3—is hidden this evening's top prize: the fanciest car in the United States of America. Behind each of the other two doors is a goat. Lucy, I don't think that you want the goat." Prompted by the show's directors, the audience laughs at this. "What do you think, Lucy—which door is the car hidden behind?"

"I don't know . . ." begins Lucy, whispering into her microphone, and then, after a few intent moments, she says: "I think that the car is behind the middle door, the one with the number 2."

"Lucy, you've chosen door number 2. Have you considered that carefully?" This is a completely pointless question, since there is nothing to consider. Monty Hall merely wants to unsettle his candidate a little. "Lucy, I know of course which door the car is behind," he says. "But I am not allowed to tell you, alas. The management will not permit it, although I would love to, since you are so nice. But I can help you a little. Have a look at what is hidden behind door number 3."

Monty Hall goes to the right-hand door with the number 3 and opens it, revealing a goat.

"Door number 3 would have been a bad choice. Thank God you didn't choose it!" Monty Hall gives Lucy a charming smile. "The car may indeed be behind door number 2."

"Or behind the left-hand door with the number 1," Lucy points out.

"Indeed," agrees Monty Hall. "And that's why I'm going to give you another chance. Before, you chose door number 2. Now you have the chance to switch and choose door number 1. What's your decision, Lucy?"

Clearly, there are four possibilities here. One: Lucy insists on door number 2, Monty Hall opens it and behind it is the brand spanking new car. Cue huge applause from the audience. Two: Lucy insists on door number 2, Monty Hall opens it and reveals a bleating goat, prompting a chorus of disappointment from the audience. Three: Lucy switches to door number 1, Monty Hall opens it, and the switch turns out to be a good decision, since the car is revealed in all its glory. Four: Lucy switches to door number 1, Monty Hall opens it and shows Lucy a goat. Lucy is annoyed and upset, since she previously chose the middle door with the car behind it.

All four of these scenarios occurred on numerous occasions in the hundreds of "three-door games" that Monty Hall had the pleasure of hosting, which is why Craig F. Whitaker addressed the following question to *Parade* and Marilyn vos Savant: "Suppose you're on a game show, and you're given the choice of three doors. Behind one door is a car; behind the others, goats. You pick a door, say #1, and the host, who knows what's behind the doors, opens another door, say #3, which has a goat. He then says to you, 'Do you want to pick door #2?' Is it to your advantage to switch your choice?"[1] And vos Savant's response is short and to the point: "You should switch."

She was immediately deluged by a veritable flood of readers' letters. The following three quotes come from scientists at universities and academies:

> "You blew it! . . . As a professional mathematician, I'm very concerned with the general public's lack of mathematical skills.

Please help by confessing your error and in the future being more careful."

"There is enough mathematical illiteracy in the world, and we don't need the world's highest IQ propagating more. Shame!"

"Your answer to the question is in error. But if it is any consolation, many of my academic colleagues have also been stumped by this problem."[2]

When vos Savant stuck to her guns in another article, *Parade* continued to be swamped by letters. The British mathematician Ian Stewart collected some of the most striking examples,[3] including those written by readers with the following types of character:

Apodictic: "Your answer is clearly at odds with the truth."

Conciliatory: "May I suggest that you obtain and refer to a standard textbook on probability before you try to answer a question of this type again?"

Submissive to authority: "How many irate mathematicians are needed to get you to change your mind?"

Democratic: "I am in shock that after being corrected by at least three mathematicians, you still do not see your mistake."

Macho: "Maybe women look at math problems differently than men."

Patriotic: "You made a mistake, but look at the positive side. If all those PhD's were wrong, the country would be in some very serious trouble."

We can see the editor of *Parade* entering vos Savant's office with a laundry basket full of letters. "The next load of letters in which people claim that you made a mistake. And I have to admit, I myself think you did, too. For the two closed doors, the probability is fifty-fifty that the car is hidden behind one or the other."

"I'm still right, though, because you have to take into account what happened before."

"But we are constantly taught that prior events play no part in probability. If in roulette the ball lands on red ten times in a row, the probability of red or black in the next spin is still fifty-fifty. Why should it be any different with the doors?"

"Let me explain it to you. Let's assume that the candidate—I'll call her Lucy—first pointed to the door with the number 2. How great is the probability that the car is behind this door?"

"A third, obviously."

"Correct. Therefore, the probability that the car is NOT behind the door with the number 2 is two-thirds. I repeat: there is a probability of two-thirds that the car is behind either the door with the number 1 or the door with the number 3. The door with the number 3 is taken out of the equation, because Monty Hall opens it and shows Lucy the goat behind it. Therefore, there is a probability of two-thirds that the car is behind the door with the number 1. This probability is twice as much as the probability that the car is behind the door with the number 2. There you go!"

"That was too quick for me. You're like a magician and I can't follow your trick."

"But it's obvious! I'll give you a more extreme example: imagine that Monty Hall is standing in front of not three but 1001 doors, numbered from 0 to 1000. Behind one of the 1001 doors is the car, behind each of the other 1000 doors is a goat. Say a number between 0 and 1000, which is the number of the door behind which the car is hidden."

"I've no idea—I'll have to guess, of course. I'll say . . . I'll say 729."

"But you're not convinced that 729 is the correct choice, are you?"

"Not in the least. The probability that 729 is correct is one in 1001."

"And now imagine that I am Monty Hall and I know where the car is concealed. I open 999 doors. Behind each one, you can see a goat bleating away. I only leave the doors with the number 729—your number—and 313 shut. What do you say now? Wouldn't it be wise to switch?"

"I understand," says the editor, visibly delighted. "Of course I would switch in this case. I would be mad to insist on the unlikely door number 729."

"It is a thousand times more likely that the car is behind the door with the number 313 than behind the door with the number 729," agrees vos Savant.

The editor stares at her for a few moments, as though hypnotized, then he gives a mischievous laugh and goes on the attack again: "There is one argument, though, that destroys all of your reasoning. Imagine, Marilyn, that I am a television viewer who likes watching the show *Let's Make a Deal.* Unfortunately, I turn on the TV too late and am only in time to see the scene where Lucy is standing in front of the doors with the numbers 1 and 2. The door with the number 3, with a goat behind it, is already open. If I see this image, then the probability that the car is behind door number 1 is exactly the same as the probability that it is behind door number 2. Both probabilities amount to 50 percent."

"If that's what you see, then it is true what you say," admits vos Savant.

"But now you are contradicting what you explained to me before," says the editor triumphantly. "Because then it is irrelevant whether Lucy sticks to her original choice or switches."

"Not at all," counters vos Savant, "If you can only see the image that you described to me, you can merely see the two closed doors. But you don't know what 'switch' means for Lucy, since you don't know which door Lucy chose at the beginning. As soon as you learn, however, that Lucy guessed door number 2 to begin with, the probability that the car is behind this door is immediately reduced from 50 percent to a third, that is, about 33 percent."

"You mean, this information alone changes the probability?"

"Precisely. And the more drastically the information changes the probability, the more valuable it is. Look at the example with the 1001 doors and imagine that you turned on the TV too late again and see 999 doors already open, with 999 goats frolicking around behind them. Only the doors with the numbers 313 and 729 are still closed and behind one of these is a car. You don't know anything else. Based on your knowledge of the situation, the probabilities that the car is behind either the one or the other closed door are the same—both are 50 percent. But if you know that the candidate of the '1001-door game' previously chose door number 729, then the two probabilities undergo a drastic change. There is only a probability of 1 in 1001, so practically 0 percent, that the car is behind the door with the number 729, and a probability of 1000 in 1001, so practically 100 percent, that the car is behind the door with the number 313."

Vos Savant's explanations of the three-door game demonstrate how important information is for a player who has to decide which move to make next. When there is a great deal at stake in the game, people are prepared to pay a lot of money for information, though only for valuable information. Information is of value if, firstly, it enables one to calculate the probability of such and such a move leading to one winning the game, and secondly—in the case of two possible moves—if the probability calculated for one of the moves is significantly greater than for the other one.

This is why casinos—seemingly as a generous service for the customers—put up screens by the roulette tables, showing the numbers that the ball has landed on in the previous spins. The information thus provided for the players is free—and it is indeed completely worthless. No experienced player will pay any attention to the screens, because he knows that the previous spins have no influence at all on subsequent ones. Yet naive players believe that the information is of value—and play accordingly. On the other hand, there used to be "wheel-watchers," who would intently observe the spinning of the roulette wheel. When they noticed that a small imbalance in the wheel would favor a certain section of the wheel, this was indeed a truly valuable piece of information. These days, however, the mechanism of the wheels on the roulette tables is so sophisticated that an imbalance would practically never be discernible.

Bidding in bridge is essentially nothing other than the announcement of information. In tarock, which is much more varied than bridge, bidding serves on the one hand to establish the play, but, of course, as in bridge, the other players can also glean information from the bid about the hand held by the bidder.

It is not only in games in the usual sense that information can apparently or actually be of value, however, but also in games in the broader sense of the term—in business competition or the field of politics. Entire armies of well-paid spin doctors toil away day and night to gather information, check its veracity, and analyze its value, or spread information themselves around the world, sometimes in official press conferences, sometimes as rumors that go from ear to ear and are, precisely for that reason, deemed to be of particular importance. In this way, a very special kind of game arises: the game with information. The editors and publishers of newspapers and magazines and the heads of the electronic media play this game before the eyes of the public, while the intelligence agencies, who, if spy novels are to be believed, employ faceless individuals in gray suits and trench coats, who once searched for dead drops and now strive to trace every electronic data track in our completely interconnected world—they play this game in secret. Not for nothing do they flippantly refer to espionage as "The Big Game"—as though it were a game of poker. Or, as the British intelligence officer called it during the conflict between Britain and Russia in the nineteenth century, "The Great Game."

Nobody should be under the illusion of being constantly able to fold in this game with information, of being able to escape it entirely. No matter how much data security the suppliers of the communication network promise, everything that one saves in any electronic device can be decoded in a flash by those with the necessary skills and motivation—everything. Of course, we shouldn't attach too much importance to our own sharing of electronic information. In view of the huge number of data transmissions, there is a probability bordering on nil that our own messages that concern only ourselves will be fished out of the myriads of others in this infinite flood of data.

But should a key word like "weapon" or "terror" be uttered, which the recognition systems are designed to pick up, then the message is guaranteed to end up in the agencies' channels for further inspection and, somewhere on the immense chessboard of bits and bytes, a tiny pawn will be pushed one square forward.

"Have a look at what one reader would like to tell you, Marilyn," says the editor of *Parade* and passes a handwritten letter from the laundry basket to vos Savant.

"Dear Marilyn," it says, "I am disappointed by your answer. I was once a candidate with Monty Hall and he challenged me to the 'three-door game.' After I had chosen a door and he had opened another door, behind which there was a goat, I remained true to my initial choice. You see, I could hear, very quietly through the other still-closed door, the bleating of the goat. I didn't switch and I won. What do you say to that?"

"The information he had was even better than what I could give him," is vos Savant's laconic response.

15

PLAYING WITH LANGUAGE

CAMBRIDGE, BETWEEN 1928 AND 1946

"How wonderful! God has arrived!"

John Maynard Keynes is positively glowing. "He came on the 5:15 train. I collected him at the station. He intends to remain in Cambridge." Keynes, one of Ludwig Wittgenstein's most ardent admirers, is beside himself with joy when he enters his club at the end of January 1929 with news of Wittgenstein's arrival from far-off Vienna. Nobody had expected that Wittgenstein would ever return to England after hiding himself away in Austria after the publication of his *Tractatus*, and the wildest rumors about this peculiar man had spread at the club where Cambridge professors met up.

"He lives like a recluse in villages south of Vienna, out in the middle of nowhere." "He has become a teacher of farm children and is even supposed to have written a German dictionary for them." "He beats the children at school and doesn't let them go home, even though they are supposed to work on their farms. The headmasters complain about him and he has to move from one backwater to the next." "He isn't a teacher any more. I heard from one visiting Austrian professor that he had been fired, and from another that he had voluntarily quit." "Now, he is apparently working in a monastery as a gardener." "No, he is dwelling alone in a hut in Norway, living the life of a hermit."

"And his rich relations? Doesn't anybody take care of him?" a concerned member of the club would ask.

"His rich sister Margaret, who married into the Stonborough family, got him to help design a house in Vienna. An ugly box in the style of that dreadful pederast Loos. Apparently, Wittgenstein made the architect Engelmann livid with his pedantry. With the construction workers, he was exactly as we found him to be in England before: so sure of himself that he stubbornly refused to listen to them. And he has remained as cranky as ever. He didn't earn anything from designing the house. He lives in a cell and only meets academics of our kind if his sister tempts him out of hiding. In truth, however, he doesn't want to have anything more to do with philosophy."

"When you recall how highly Russell thought of the man! He told Wittgenstein's sister Hermine that her brother would give philosophy a completely new impetus. What a bitter loss! What a waste of talent!"

When these words of disappointment were uttered in March 1928, none of those present at the club had any idea that, at that moment, Wittgenstein was listening to Brouwer's lecture at the University of Vienna's mathematical institute, a lecture that would fundamentally change his life. We can only guess what might have caused Wittgenstein to rethink. He was certainly challenged by Brouwer's unusual idea that the point of departure for all cognition was the "two-ity," the emanation of the number two from the number one, and that, from the "two-ity," the infinite mass of all other numbers arise, as, eventually, does mathematics, and with it everything that can be rationally comprehended. When Wittgenstein was fascinated by an idea, he immediately attempted to contradict it. For that alone causes the movement of our thought to come about. A fascinating idea may come close to the truth, but we come even closer if we consider what insight can be gained from the contradiction of this idea.

Wittgenstein's counter-thesis to Brouwer's idea is that it is not numbers, as Brouwer believes, but rather language that gives thought about the world its foundation. By confronting Brouwer's thinking, he comes to realize that he did not give enough consideration to language in the *Tractatus.*

Intuitively, Wittgenstein recognizes that Vienna, staggering as it is toward an uncertain political future, is not a place where he can think seriously about language in uninterrupted peace. Then he recalls the invitation extended by his English friend Frank Plumpton Ramsey—who even visited him while he was teaching in Puchberg am Schneeberg—to go to Cambridge. He remembers the famous university town that he visited just before the World War and where he made the acquaintance of Bertrand Russell and the members of the intellectual society known as the Cambridge Apostles. And he senses that it is at Cambridge University that he might well find the repose required for his deliberations.

"You are most welcome!" his teacher, mentor, and—if such a thing can ever be said with Wittgenstein—his friend Bertrand Russell greets him. "We'll organize a position for you at the university."

"There is just one small problem," interrupts the man sitting next to Russell. "Mr. Wittgenstein cannot teach here. He is a qualified engineer but has no doctorate."

Russell is not put off by this. "We can easily put that right," he says. "We'll take one of Wittgenstein's papers and make it his dissertation. Moore and I will conduct the final exam."

"But I haven't written anything since the *Tractatus* that could be used for a dissertation," objects Wittgenstein.

"Then we'll simply take the *Tractatus* itself," decides Russell.

No sooner said than done: the *Tractatus logico-philosophicus* is accepted as Ludwig Wittgenstein's doctoral thesis in Cambridge, and he is called to defend his work before Bertrand Russell and George Edward Moore. It goes without saying that this "exam" about a book that both examiners praise as a milestone in twentieth-century philosophy is a mere formality. At the end of the exam, Wittgenstein takes his slim volume, stands up, claps the two examiners on the shoulder, and says to them as he leaves, "Don't worry—I know you'll never understand it."

In his examiner's report for Wittgenstein's "dissertation," Moore writes, "I myself consider that this is a work of genius; but, even if I am completely mistaken and it is nothing of the sort, it is well above the standard required for the PhD degree."[1] When Moore retires ten years later, Wittgenstein takes over his chair and extols the virtues of his predecessor, because a single sentence of his had a decisive influence on Wittgenstein's thinking about language. Moore's sentence is:

"It's raining but I believe that it is not raining."

Wittgenstein calls this sentence "Moore's paradox," and he likes to talk about it in the seminars that he holds in Cambridge.

"The sentence 'It's raining but I believe that it is not raining' represents the most important contribution that my predecessor George Edward Moore made to philosophy. Let me explain why.

"On the one hand, the sentence sounds absurd. And it is indeed absurd. On the other hand, it can be true. For it consists of two parts that, factually, have nothing to do with each other. The first part is the assertion that it is raining. This is a meteorological statement about the external world. The second part is an assertion about what I believe. This is a psychological statement about my internal world. Both assertions are taken from completely separate domains. How can it therefore be possible that they contradict each other? It is simply impossible. A formal contradiction can thus be ruled out.

"And yet I am talking absurdly when I say: 'It is raining but I believe that it is not raining.' It's absurd because *I* am the one saying it. It is not *the* language that brings about the contradiction, it is *my* language that makes the sentence an absurdity.

"And because the sentence is absurd, I realize that there is no language *per se*; there is only *my* language, the language of the person who is speaking.

"How is it that you understand my language? How is it possible that I can understand you when you ask me a question in your language? The answer is that it is only possible because we have agreed upon the same language game.

"It is indeed like in games: if a chessboard is lying on the table and the pieces are scattered next to it, that is not a game of chess. It may have been a game of chess before, and the players have removed the pieces from the board. Or it may become a game of chess at some point, when the players have organized the pieces on the board according to the rules of chess and then take turns to make their moves. But with the pieces scattered around as I can see them, it is no more a game of chess than a system of words that nobody is speaking or writing down in a sentence. My language is like a game. It is only when I am playing the language game that it comes into being. It is a deadly serious game—we battle using language, and we battle with language."

Wittgenstein pauses in this explanation that he accompanies with much gesticulation, looks around, and senses many pairs of uncomprehending eyes directed at him.

He does not know if anybody at all in the seminar has understood what he has said. He doesn't even know if he recognizes himself in his own words; whether his thinking is truly what he expresses with his propositions. Perhaps he has to say it even better. He jots down notes upon notes, fragments of thoughts, rather than self-contained theories. Deep down, he knows that every theory remains a patchwork and can never provide the ultimate explanation of the existence in which he finds himself. He knew that already when he wrote the *Tractatus*, in which the penultimate proposition, number 6.54, reads:

> My propositions are elucidatory in this way: he who understands me finally recognizes them as senseless, when he has climbed out through them, on them, over them. (He must so to speak throw away the ladder, after he has climbed up on it.)
>
> He must surmount these propositions; then he sees the world rightly.[2]

After completing the *Tractatus*, Wittgenstein wanted to rid himself of his addiction to philosophy. But after experiencing Brouwer's lecture in Vienna, he remains under its spell for the rest of his life. It is like an addiction to games, the addiction to the language game.

Wittgenstein is convinced that "Philosophy is a battle against the bewitchment of our intelligence by means of language."[3] All nonsense that arises has its origin in the breaking of the language game's rules. Wittgenstein experiences actual physical discomfort to see people breaking the rules of the game and imagining something in this, indeed believing that they have gained insights or truths from it. One such "dimwit," as Wittgenstein derogatorily refers to him, Karl Popper, is invited to give a talk in Cambridge in 1946, and Richard Bevan Braithwaite manages to convince Wittgenstein to meet Popper at a gathering of the Cambridge Moral Sciences Club.

Braithwaite is a thinker who comes from the field of mathematics, who is interested in the philosophical foundations of probability theory, and who is one of the first to suggest using game theory to choose the best hypotheses to cope with ethical or religious conundrums. He himself likes playing games, too, in that he takes pleasure in egging on two opponents and observing how they deal with the conflicts that thus arise.

Braithwaite knows that both Wittgenstein and Popper find it difficult to remain calm when they hear assertions that they consider to be nonsense. He knows that, while both men are from Vienna and have Jewish ancestry, their common features barely unite them. The differences between the two are patently obvious: on the one hand, there is the graceful Wittgenstein, the masterful, all-knowing genius from the wealthiest background, always with a troop of young men dressed like him in tow. On the other hand, there is Popper, thirteen years Wittgenstein's junior, but much stockier, an upstart, ambitious, vain, and resentful. Wittgenstein once so airily declined to join the Vienna Circle, whereas Popper would have been only too glad to become a member—except that he was never asked.

With fiendish delight, Braithwaite is looking forward to seeing the two men come up against each other. What is more, along with many other dons and students, Bertrand Russell is also coming to the shabby, unheated Room 3, in Staircase H of the Gibbs Building. For Braithwaite, Russell's presence adds particular spice to the whole thing, since Russell has increasingly distanced himself from his old pupil and friend Wittgenstein, ever since the latter had focused his thinking on the language game, and has indeed shown an interest in Popper, who, from his exile in distant New Zealand, wrote papers on economic and social theory that Russell considers to be of value.

Wrapped in coats or gowns, the attendees are sitting packed together in the clammy room. A small fire is burning in the hearth, but its heat has little effect on this cold October day. Wittgenstein, who is chairing the meeting, and Popper, as the guest speaker, have taken their places in front of the fire and, after a brief, informal greeting, Popper is invited to speak about the topic he has chosen, "Are there philosophical problems?"

During the first few minutes, while Popper is declaiming his prepared text in a slow, monotonous voice, Wittgenstein listlessly digs away at the weak flames with a poker, seemingly absentminded. Suddenly, he turns around and interrupts the speaker: "Popper, this won't do. What you are presenting to us here is gibberish, claptrap—absolute nonsense. Do you have any idea at all what philosophy is?" Moments of shocked silence from those present follow, before Wittgenstein continues, the poker raised in his hand like a conductor's baton: "The aim of philosophy is to show the fly the way out of the jar. And what are you doing? You're pouring sugared water into the jar!"

"I don't understand . . ." begins Popper, but he doesn't get far.

"You simply have no command of the language game," rebukes Wittgenstein. "You treat the rules of language like a chess player who moves the rook diagonally and uses the queen to jump like a knight. But with your meaningless talk, you are simply leading us, and finally also yourself, down a dead end: the problems of philosophy which you are blustering about are pure humbug."

"Why do you talk about humbug, here in the land of Berkeley and Hume, when I raise the question as to whether we perceive the world via our senses?" says Popper, trying to defend himself.

"Then you should concern yourself firstly with psychology, which will explain to you how we can speak reasonably and usefully about the senses. Once you have learned even just a little about that, you will see that your problems will vanish into thin air."

"Or the fundamental problems of mathematics, about which even the experts are at loggerheads. For instance, the question whether there is really infinity, indeed whether there are in fact several infinities," retorts Popper, sticking to his guns.

"There is, there are!" jeers Wittgenstein. "Do you know what 'there is' actually means? When I was having breakfast in the dining hall today, the waitress said to me: 'There is no marmalade today.' 'No,' I answered, 'There is always marmalade. Even today there is marmalade.' She obviously didn't understand my little joke. After all, she cannot know that she and I were using the words 'there is' in language games with completely different rules. Did you know that one can write a serious philosophical paper using jokes alone, without descending into farce? But you don't have enough of a sense of humor for that, Popper. The waitress did laugh—perhaps she understood me a little after all."

Before Popper can counterattack, Russell intervenes. "You're confusing everything, Wittgenstein, you're confusing everything," he complains. But Popper will not be restrained. "Trying to reduce everything to language games completely ignores reality," he says. "Why should I care about the rules of the language game when ethics is all about the rules of coexistence? Or do you wish to deny, Wittgenstein, here in the Moral Sciences Club, that philosophy is concerned with the validity of moral rules?"

Now facing Russell rather than Popper, Wittgenstein says heatedly, "It was always clear to us, Russell, that philosophers fall silent when they are expected to pass judgment on whether thought, speech, or action is good or evil. You yourself were always convinced that moral insights could not be considered to be 'knowledge'. And now this narrow-minded fellow from no man's land"—Wittgenstein gesticulates wildly with the poker in Popper's direction—"comes and speaks about the 'validity of moral rules'! Ridiculous! He isn't even in a position to give a single example of a moral rule!" He leaps up, throws the poker down on the hearth, strides to the door, and slams it shut behind him.

For several seconds, there is a dead silence in the room. Then Russell interrupts it in a humorous voice: “Perhaps there is at least one moral rule—one shouldn’t threaten a guest speaker with a poker!”

If truth be told, nobody knows whether the dispute between Popper and Wittgenstein on that evening of October 25, 1946, truly happened like this or not. It was only decades later that living witnesses of the event were asked about it, and their answers differed greatly. Popper, who was always out to present himself in the best possible light and dismiss others—up to and including Einstein and Bohr—as obsequious stooges of his theories, which he himself considered to be so brilliant, claimed that it was he who, in response to Wittgenstein’s demand to name an example of a moral rule, had readily replied, “One shouldn’t threaten a guest speaker with a poker.” To which, according to Popper, Wittgenstein could think of nothing to say and so left the room in a rage. Popper, a notoriously bad loser, was always able to twist stories to ensure he came out of them on top.

16

PLAYING WITH EMOTIONS

IOS, AROUND 850 BCE; BARCELONA, 2014; ROME, 1900; VIENNA, 1786

"All the world's a stage."

In William Shakespeare's *As You Like It*, the melancholy Jaques sits at the table of the exiled Duke Senior and utters these famous lines about life and the world: "All the world's a stage, and all the men and women merely players; they have their exits and their entrances."

Alexander Mehlmann, a game theorist at the Technical University of Vienna, is fond of illustrating the manifold games of existence with examples taken from literature. This is in no small part down to the fact that he himself is not only a mathematician, but also a poet.

One such example is taken from the *Cypria*, an epic poem that, by some researchers, is attributed to Homer's relative Stasinus. It contains a paraphrase that shows how a "move" by one of the "players"—in this case, the crafty Odysseus, who wishes to escape being enlisted into the army against Troy—is thwarted by the "move" of his equally cunning opponent Palamedes, the man sent to fetch Odysseus by the Greek leader Agamemnon. But let us hand over to Alexander Mehlmann himself:

> Only a few seemed not to hear the call to arms. One of these, Odysseus, had in fact left three urgent messages from the Achaean general staff unanswered. According to rumor, the Delphic Oracle

had prophesied that should Odysseus participate in the war, he would be condemned to twenty years in exile. Now Agamemnon, his designated commander, had arrived in Ithaca to apprise, in person, him who was liable for military service of the general mobilization.

The Mycenaean delegation found the island in a desolate state. An unusually fine vintage was rotting on the vines. The royal palace on the mountain Aetos sheltered only its servants. Descending along the western slope, the armed troops were met by Penelope, wife of Odysseus, distraught, in tears, and bearing her infant son Telemachus in her arms.

"Where is your husband, woman?" barked Agamemnon. The queen of Ithaca indicated with a nod of her head the path to the lonely beach below, where a powerful form was plowing meandering furrows into the soft sand. Horse and ox were yoked to the plow. The man who was plowing wore a peaked cap and without surcease was sowing salt. It was Odysseus, whom the gods had apparently struck with madness.

"Unfit for military service," observed Agamemnon unhappily. Then Palamedes grasped the infant and placed him in front of the plowshare. Would Odysseus run the plow over his son?[1]

Odysseus now had to decide between two "moves": either he could continue to pretend to be insane and thus save twenty years of his life but sacrifice the life of his newborn son, or he could steer the plow away from the child. He spontaneously opted for the latter. In his turn, Agamemnon had to decide whether this swerve by Odysseus was merely down to chance and Odysseus was indeed insane or whether, by sparing his son, Odysseus had thus unwillingly exposed his own trickery. Agamemnon, too, opted for the second choice, and Odysseus was forced to join the Greek warriors. Palamedes, however, who had created the dilemma, had not reckoned with Odysseus's implacable desire for revenge, which would later bring about Palamedes's own downfall in a similarly cruel "game."

Even the Ancient Greeks were already aware that games become truly interesting when they are made up of many different rounds and one never knows which one will be the last.

All the world's an epic, a drama, a play, and the limits of this world are simultaneously the limits of language. Seen in this way, Ludwig Wittgenstein's language game gains an existential depth on the stage of the world.

The extent to which this holds true in the subtlest nuances of language can be seen from a remarkable experiment conducted by the psychologist Albert Costa from the University of Barcelona, who confronted test subjects with the following dilemma, presented in different languages:

> A train is hurtling straight toward five people and you have the chance to prevent the calamity by pushing a large man onto the tracks. You thus sacrifice one life, and save five others. What is your decision?[2]

The question contains a dilemma that the Ancient Greeks had already brought into play. Either one decides that five lives are worth more than one, which would be the position of utilitarianism, which considers the maximum utility in questions of ethics. For the large man, this choice would mean certain death. Or one makes the decision that one cannot, under any circumstances, kill an innocent person—one's conscience forbids it and this is true regardless of the consequences, which one personally can do nothing about.

The fascinating result of Costa's experiment was that it makes a difference whether the moral dilemma just described is presented to people in their mother tongue or in a foreign language. When Costa's Spanish test subjects had to judge whether it was right to sacrifice a "hombre grande," they were less willing to do so than they were with a "large man." It does seem as though our decisions are fundamentally dependent on whether we consider them in our mother tongue, in the language game that has become second nature to us, or in a foreign language, an acquired and unfamiliar language game. Our mother tongue inspires our fantasy and imaginative power, inducing strategies that are not based on a cost-benefit calculation, but rather originate from other, deeper sources.

We are gripped still more by the drama if the language is underscored with music. While we view the prisoners' dilemma as told by Albert Tucker as an intellectual exercise, it becomes a stirring tragedy in Puccini's *Tosca.* Alexander Mehlmann recounts the story with subtle humor:

> In a desperate attempt to save her lover Cavaradossi, who has been condemned to die before a firing squad, Tosca makes a fatal pact with the henchman Scarpia. She is prepared to give herself to him if he agrees, before the encounter, that the execution will be carried out with blank cartridges. The Charybdis of a prisoner's dilemma draws Tosca and Scarpia into its vortex, not, of course, without giving each of them the opportunity to sing a final aria. Each of them breaks the agreement by choosing his or her strictly dominant strategy. Thus, Scarpia secretly countermands the order to switch to blanks, while Tosca, for her part, stabs her love-crazed adversary with a knife that seems somehow to have been left on stage by an absent-minded property mistress.[3]

There are two elements that differentiate the game played on the stage of the world from the games that we have looked at so far:

First, this game is unique and cannot be repeated. No player can therefore make use of a calculated probability of the game ending to his advantage. It is as though one were only permitted to roll a die once. If one throws a six, as desired, then one can be justifiably pleased, since it is far more likely not to throw a six. One can be even more pleased if one bets on zero for a single spin of the roulette wheel and the ball actually does land in the slot marked zero, for such an occurrence is even less likely than throwing a six with a die. But that is as far as it goes.

Secondly, this game is shot through with seemingly random factors, which are actually not so incidental after all, because they are the elements that address the emotions. The purely rational player, whose decisions are based on a table of sober calculations, is out of place in the one-off game on the world stage that is of critical importance for one's very existence. Those who disregard the irrational, spontaneous, and impulsive aspects of life overlook a very fundamental factor.

The difference between the rational treatment of a game on the one hand and its artful realization on the other hand is particularly noticeable in the operas written by the great composers, where what happens on stage goes straight to the heart because of the music, although the plot itself is usually based on bland or sober texts by the librettists.

Take Mozart's opera *The Marriage of Figaro*, for example:

"He is writing an opera based on that detestable Beaumarchais play?" asks music director Franz Orsini-Rosenberg of his informant Antonio Salieri, if we are to believe Peter Shaffer's play *Amadeus.* In the years before the French Revolution, Beaumarchais's play *La Folle Journée: ou le Mariage de Figaro* was generally considered to be a highly provocative dig at the aristocracy.

"It's a French play. It has been banned by the emperor," explains Orsini-Rosenberg to the simple-minded Kapellmeister Giuseppe Bonno, who is sitting next to him, and he makes sure that Emperor Joseph is immediately informed of Mozart's wicked intentions.

"Mozart," begins the emperor in a stern voice, when the composer is summoned to appear before him, in the presence of Bonno, Orsini-Rosenberg, the court librarian Gottfried van Swieten, the court chamberlain Count Johann von Strack, and other senior court officials, "are you aware that I have declared the French play of Figaro unsuitable for our theaters?"

"Yes, Your Majesty."

"Yet we hear you are making an opera from it. Is this true?"

Mozart looks around in bewilderment. One of those present must be an intriguer who has found out and revealed the plan he, Mozart, has secretly hatched with Lorenzo da Ponte: "Well, yes, I admit it is."

The emperor's voice rises noticeably: "Mozart, I am a tolerant man. I have largely abolished censorship in my realm, although that has brought me little gratitude from my subjects. But I cannot and will not allow pernicious plays to be performed in my theaters. This Beaumarchais and his *Mad Day* have put nothing but mischief into people's heads in France. The play has sown seeds of discord and stirred up the lower classes against the aristocracy. My sister Antoinette writes to me that she is beginning to be frightened of her own people."

"I know nothing of that, Your Majesty," protests Mozart. "I see in the play only a comedy, a harmless farce."

"A farce! How naive you are!" The emperor slams his hand down on the arm of his chair. "Strack, bring me the libretto!" With astonishment, Mozart sees that the opera's entire libretto is already doing the rounds at the Imperial Court. But he has little time for surprise, since the emperor begins to read out the text without further ado: "'Se vuol ballare, signor contino'—'If you want to dance, Mr. Count'—that's how the insolent servant commences, challenging a count? As though he were on an equal footing with him! Strack, what do you say to that?" Without waiting for an answer, however, the emperor continues reprovingly: "Did you not think anything of it, Mozart, when you read these lines? Do you not see that the play gives the people carte blanche to rise up in revolt, and robs the aristocracy of their dignity?"

"But Your Majesty, there is no trace of a revolt against the aristocracy to be found in my music. I am incapable of writing music that would inspire such an uprising."

"You heard His Majesty's words, Mozart," interrupts Orsini-Rosenberg officiously. "Kindly forget the piece."

"But I have almost finished composing the opera," Mozart protests in reply, and then turns pleadingly to the emperor: "Permit me, Your Majesty, to play just this one aria on the piano—just this one cavatina!" Without waiting for a response, he hastens to the instrument and plays Figaro's first-act aria to those gathered in the council room. He breaks off just before the end, rises and turns to the emperor again: "Can Your Majesty hear even a single note of revolt and insubordination? Your Majesty cannot have done, because I have composed all that away. Nothing of all the pernicious things that might be deduced from the text can be found in the aria. What remains is the dramatic element of the play—what remains is the music . . . my music. My music overrides everything; my music mitigates everything—it casts a spell and makes us forget all politics."

"You are very sure of yourself," says the emperor, a little more graciously. "You must answer to my librarian. If van Swieten's opinion of your opera is favorable, then I will allow it to be performed."

Mozart breathes a sigh of relief. He knows Gottfried van Swieten to be a sympathetic supporter of his, the two of them bound by a shared devotion to Johann Sebastian Bach's music. Yet when van Swieten later conducts a confidential interview with Mozart, in accordance with the emperor's instructions, he begins with a skeptical question: "Why such a vulgar subject matter, Mozart? This farce is not worthy of your art."

"Because it is the most wonderful game that is being described here. It wasn't da Ponte who dug out this Beaumarchais play. I myself found it and I fell in love with it straightaway. I was captivated by the game with the characters, the game with the emotions.

"Note that it is neither Count Almaviva nor his wife who pulls the strings of this game, and not even Figaro, either, despite his challenge to the count. No, it is actually Susanna, this almost omnipresent and truly splendid character. She plans everything, is much cleverer and more calculating than her betrothed, and she leaves the audience in no doubt as to her true intentions. She directs the game magnificently. And how wonderfully she is able to hold the count like a puppet in her hands, although he is under the illusion that he can master her."

"But my dear Mozart, you portray the count as a particularly unpleasant fellow, as a spoilt, bad-tempered, and domineering man. Is that really necessary? That is the politically sensitive element that concerns His Majesty."

"Believe me, my dear van Swieten, I beseech you: you will ignore all of that when he sings. And the play's dramatic structure demands this portrayal. You see, the climax of the whole game comes when the count seems to end up as the loser and yet manages to win back everything, when he is driven to remorse in the finale of the last act when faced with his wife dressed as her maid. You will never forget the way 'Contessa, perdono' leaps up a sixth or how the following 'Perdono, perdono' leaps up a seventh, so heart-rending is the count's plea for forgiveness."

"But nobody will really believe that the count will suddenly change his character, even if he does recognize that he went too far that day."

"Dear van Swieten," Mozart beseeches his benefactor, "that is completely irrelevant. The whole thing is just a game. The count does not really exist—he is merely a character in the game. When the curtain falls, everybody will call out 'La commedia è finita!' The only thing that counts for me is that my music overrides all doubts. In a spoken play, perhaps the audience does get mixed up between what happens on stage and reality. In a play, His Majesty may be right in saying that what is acted out on stage can be confused with real life. But in an opera, with music . . . ! With music, a count can treat the common people as condescendingly as he likes and this does not cause a revolution; instead, the listener delights in a perfect harmony that goes beyond politics."

If Mozart's music succeeds in elevating the action of *Marriage of Figaro* above a cynical and political level, it also manages to achieve a similar disassociation in the opera *Così fan tutte.*

In terms of its structure, this work is a curious chamber play. Fiordiligi and Guglielmo are introduced at the beginning as one amorous couple, with Dorabella and Ferrando as the other. They are joined by another pair of characters, Despina and Don Alonso, who, in the course of the opera, succeed in making the two ladies, Fiordiligi and Dorabella, forget their devotion to their original lovers and each fall in love with the other man, before finally, serene and sober again at the end, returning to their original betrothed. These days, out of a misplaced respect for Mozart, people zealously suppress the criticism that the opera actually depicts a frivolous and immoral game. In the nineteenth century, on the other hand, they were unashamed to point this out and, in doing so, were much more honest than we are today, for the text of *Così fan tutte* truly is frivolous and immoral. Despina and Don Alonso, who pull the strings throughout the whole game, are malicious and mocking cynics, whose experience of life has shattered beyond repair any belief in love and loyalty they might once have held. Indeed, criticism of the opera's libretto in the nineteenth century was so extreme that *Così fan tutte* was performed in adapted versions with completely new texts.

As it happens, however, Mozart's music renders this well-meaning, though not necessarily justifiable correction superfluous. The fact that *Così fan tutte* is performed today with Lorenzo da Ponte's original libretto does not mean that the text has suddenly shed its immoral frivolity, but is rather down to the fact that the music transforms the game presented in the opera to such an extent that questions of morality are essentially dispelled.

"But in an opera, with music . . . !" Mozart might say, convincing his critics as before, ". . . with music, people can declare undying love and then betray each other after all, and yet this does not cause a moral problem; instead, the listener delights in a perfect harmony that goes beyond moral law."

17
PLAYING WITH EXISTENCE

PARIS, 1662

"You must play this game!"

Antoine Gombaud sits at Pascal's bedside and listens as his terminally ill friend addresses him with incredible strength of will: "You must place your bet! You must make a wager! You will not escape the consequences of your decision. Absolutely everything is at stake in this game. It cannot be avoided. You must play this game!"

Gombaud had not expected the conversation, which had begun so harmlessly, to take such a turn. Games had been far from his mind on his way to the house where Pascal's sister Gilberte lived with her husband Florin Périer and her critically ill brother Blaise. Gilberte Périer had written to say that her brother did not have long to live, and that Gombaud would have to visit him soon if he wanted to see him again.

"I came in one of your 'carrosses à cinq sols,'" says Gombaud, in an attempt to lighten the conversation. The "carrosses à cinq sols" were an invention by Pascal and his friend Artus Gouffier, the Duke of Roannez. The two men founded a company that enabled people to travel in a horse-drawn carriage ("carrosse") for the price of five sous (then known as "sols") from one station in Paris to another. Pascal and Roannez can thus be called the inventors of public local transport; five lines served by horse-drawn omnibuses linked various parts of Paris from March 18, 1662, onward.

"You shouldn't really know that I am linked with that enterprise," says Pascal, with a weak but dismissive gesture, as he lies sweating in his bed on that hot August day. "Actually, I wanted to have two fares: five sous as the normal price and two sous as the price for the poor. But my business partners overruled me. I can understand why—the carriages have to travel, even if only one passenger gets on. Even at a price of five sous, it doesn't really pay, and at two sous, there is the danger that, sooner or later, the company will go bankrupt."

"But you are currently earning a decent sum from it?"

"Roannez owns the lion's share of the company. I myself have little interest in what I earn from it. I used to place importance on such things, back when I constructed a calculating machine at the age of nineteen. At the time, I thought I could make a lot of money with my invention, but the high production costs didn't allow this. I'm simply not made to be a businessman. It's not important anyway. I don't need money now—you can't take anything to the grave."

Pascal is fully aware of the critical state of his health. Gombaud attempts to steer the conversation in a different direction: "But your relatives could make good use of the profits from the company. When I entered the house, I made the acquaintance of your delightful niece Marguerite."

"Marguerite is a good child. And I experienced at first hand a true miracle with her."

"What happened?"

"For years, her left eye was afflicted by a serious infection, which attacked her nasal bone and perforated her palate. The doctors were as usual at a complete loss and wanted to 'cauterize' the growth, they said. Before my sister, her husband, and I would allow such nonsense, we sent the child to Port Royal des Champs. The nuns there placed on the affected eye a thorn that is claimed to come from Jesus's crown of thorns. Whether that is true or not—the eye was healed that very evening and, when the doctors visited the child some days later, they found her in perfect health." Gombaud stares disbelievingly at Pascal. His expression suggests that he thinks Pascal is delirious with fever. But Pascal reiterates his point: "I myself was witness to this miracle. I have written a report about it and given this to a notary. I am not imagining things, my dear Gombaud. What I have told you is the complete truth."

"A miracle is the dearest child of faith," begins Gombaud, who cannot restrain his skepticism. "But in my view, miracles feed doubt about God much more than faith."

"Why is that?" asks Pascal.

"It's wonderful that your niece has been healed, and I do not begrudge her this in the slightest. But what about all the others who suffer, perhaps even more piteously, and yet no miracle occurs to heal them? Or, to put it another way, what did Lazarus gain from being raised from the dead, since he had to die again in the end?"

"A reasonable question," responds Pascal. "It is true—if a miracle happened in order to make a small improvement to the state of the world outside, it would be pointless, almost a mockery. Nothing has changed in the world as a result of this particular miracle. But something has changed in me, because I understood the miracle that I witnessed to be a sign."

"You mean the miraculous healing of your niece prompted you to change your life?"

"It strengthened the resolution that I had taken one and a half years before. This had been prompted by a curious experience, the impact of which I can barely describe in words. During the very night that it overwhelmed me, I recorded it as best I could on a piece of parchment. I keep this writing constantly at hand." With some difficulty, Pascal points toward his coat, which is hanging on a hook in the room. It is only after his death that his sister will find a thin strip of parchment sewn into the padding of the coat, the *Mémorial* of Blaise Pascal, in which he describes in halting words a mystical experience: in the night of November 23, he saw God, not the God "of the philosophers and scholars," but rather, in a reference to the biblical account of the burning bush seen by Moses, "fire." It was for him such an intense, mystical experience that he withdrew completely from Parisian society and the fashionable world in which he had previously indulged, in order to devote himself entirely to a life of piety.

"But think what you gave up!" Gombaud points out. "In the salons of the Marquise de Rambouillet or Madame de Sablé, the two of us would meet the very cream of Parisian society. I can still remember how you would embellish the conversation with your witty comments. Your departure left behind a deep void."

"I'm sure you very quickly got over it. It's very easy: one moment you are thinking of me, and then somebody throws you a ball and you have to throw it back, because you want to win the game, and already you are distracted again. It's a curious thing—there you have the king, called to guide an entire state, and he is completely consumed by a rabbit hunt.

"What is man's true calling, my dear Gombaud? You know the answer as well as I do: it is to think. That is where his dignity and merit lie. And his duty lies wholly in thinking in the right way.

"Did I do that back then? Not at all—I thought of dancing and playing the lute, of singing and writing poetry, of wrestling and other things, of fighting and playing the role of a king, without thinking what it actually means to be a king and to be a man.

"And on top of that, I was arrogant, I wanted to be known throughout the whole world, even among the people who will only be alive when I am no more. I was so vain that the respect of five or six people standing around me amused and satisfied me. Even curiosity and the thirst for knowledge are mere vanity. We usually only want to know something in order to be able to talk about it. We would never travel overseas if we couldn't talk about it. Not one of our self-satisfied scientists would investigate the world for the pleasure of seeing alone and without the prospect of ever being able to report on their findings."

"And now you think in the right way? What makes you so sure?"

"You have a gift for games, my dear Gombaud," says Pascal. "Let me explain to you using the language of games. When I made the decision to change my life, it was as though I was tossing a coin: the coin can fall on heads or tails, and I wagered everything on heads.

"I don't know how the coin will fall. It is still spinning in the air—and that is how I see my existence, too, in this strange, dismal, endless world full of mystery. I look in every direction and see all around me only darkness. Everything in nature fills me with doubt and this makes me uneasy. If I could see nothing in nature that pointed to a divine presence, I would resolve to deny the existence of God. If I could see everywhere the traces of a creator, I would be fully at peace with my belief. But since I see too much to deny and too little to be free of doubt, I find myself in a lamentable state of uncertainty.

"I even believe that the universe does not simply exist as a gigantic stage on which I, as an insignificant extra, have a tiny role to play for a fleeting moment. I believe that the situation is quite different: the entire universe, from the most distant stars to the minutest mite in my hair, exists for the sole purpose of making me aware of the extent of my helplessness. The coin is spinning in the air and neither the 'esprit de géométrie,' my logical thinking, nor the 'esprit de finesse,' my intuitive understanding, can tell me whether it will land on heads or tails.

"What would it mean if I were to bet on tails in this game? Then I would be betting on not believing, on the ultimate rejection of a redeeming, merciful presence, on the denial of God. And what does it mean that I have bet on heads? That I have bet on believing, on an Eternal One, with the sole purpose of my existence being to align my life according to him.

"'What pay-out can I expect if I win this game?' That's the question that every gambler will ask, am I not right, Monsieur Gombaud? In my wager, the answer is obvious: if you bet on tails, on not believing, and you win, then you have won literally nothing. If you bet on tails, on not believing, and you lose, then you have lost the kingdom of heaven. If, on the other hand, you bet on heads, on believing, and you lose, then you have lost literally nothing. And if you bet on heads, on believing, and you win, then you have won the entire kingdom of heaven.

"For you as a gambler, it must therefore be clear: in a wager, in a game, where, if you bet on tails on the one hand, you lose everything but can win nothing, and if you bet on heads on the other hand, you can lose nothing but win everything, then you simply must bet on heads."

"But that proves nothing."

"That proves nothing," agrees Pascal. "But that is the point on which this whole bet is based."

"And how do you know that the chances of winning are fifty-fifty, like when tossing a coin?"

"I don't know that either. But it is completely irrelevant. Even if it was like roulette and one had to bet on zero if one wanted to bet on believing, then you should do so at all costs. You have nothing to lose."

"But how do I know that the God on whom I am supposed to bet will reward me with the kingdom of heaven? Perhaps he has no interest in me at all. Perhaps it is not a benevolent God, but rather the malicious devil, tricking me out of my entire stake, who has the last word. Perhaps your wager is revealed to be a con game, the result of which is predetermined, and there is no way to escape your loss."

"Do you know what, Monsieur Gombaud? Thinking about it all isn't worth it, in my opinion, for we will never know the answer. But the game is there, waiting to be played. It has to be played, and your bet is required. That is all that matters, that alone counts."

"I don't accept that. I am only interested in games in which I can win here and now, not a game that concerns an uncertain future. In your wager, Monsieur Pascal, I'll pass. I simply won't play."

"You must play this game!"

Antoine Gombaud sits at Pascal's bedside and listens as his terminally ill friend addresses him with incredible strength of will: "You must place your bet! You must make a wager! You will not escape the consequences of your decision. Absolutely everything is at stake in this game. It cannot be avoided. You must play this game!"

After a short, exhausted pause, during which Gombaud looks awkwardly at his weary friend, Pascal continues in a calmer tone: "You believe that you only live for the here and now, for the present. You are completely wrong. Monitor your thoughts. You will realize that they are all concerned with the past and the future. When you think of the present, you do so with the aim of seeing how to master the future. The present cannot be your goal. The past and the present are purely the means—the future alone is the goal. That is why you never truly live and you merely hope to live. And because you are constantly preparing for a future happiness, it is inevitable that you will never be truly happy."

"I do not seek the ultimate happiness that you, Monsieur Pascal, have set your sights on. I merely seek a little happiness. I know that each game is followed by another. When I win one game, I am fully aware that I can lose the next. A little happiness gratifies me, not the ultimate bliss."

"If that is how you think, then you too have placed your bet in my wager. And you have bet on tails. Perhaps I too used to have this cheerful carelessness that carries you through the perils of existence, but I have now lost it. I see people as though bound in chains, all condemned to death. Every day, some of them are throttled before the eyes of others. Those who remain see their own fate in the fate of their fellow captives and they wait, when they look at each other, full of suffering and bereft of hope, until it is their turn. That is how I see the destiny of all people."

"Then games are my escape, the small, pleasurable distractions that allow me to forget the absurdity of existence that you describe for a few pleasant hours of my life. I ask for no more than that. It would be unreasonable to demand more."

But Pascal doesn't hear Gombaud's answer. "My wager is not a game like other games," we hear him murmur. "It goes far beyond that. Those who bet on heads in my game have freed themselves from the chains of existence." And we can allow his weak voice, with one final effort, to anticipate one of the Danish existentialist Søren Kierkegaard's ideas:

"Imagine a theater audience eagerly awaiting a performance. Suddenly, the theater director appears before the curtain and announces that a fire has broken out in the theater. The audience believes that his announcement is part of the play. The director realizes this and shouts in desperation that everybody should flee and the stage is already engulfed in flames. The audience cheers with delight because they think the director is playing his role so magnificently. Even when the flames lick away at the curtain, the audience's applause knows no bounds, overwhelmed as they are by what they think is the staging.

"My wager is like the announcement of the theater director. Those who bet on tails remain in their seats, those who bet on heads flee into the open air. With my wager, we are not playing out a drama, where the murdered characters smile and take a bow after the performance. With my wager, we are playing with existence."

NUMBER GAMES

EXERCISES

1. Méziriac's think-of-a-number game

In his book *Problèmes plaisants et délectables qui se font par les nombres*, published in the seventeenth century, Bachet de Méziriac presents the following game: Two players sit opposite one another. The first person names a number between 1 and 10. Then the second person thinks of a number between 1 and 10, adds it to the one just named and calls out the total. The first person then adds, in his turn, a number between 1 and 10 and calls out the new total. The two players alternate in this way until one of them is able to call out a number greater than 100. The one who can say a number greater than 100 has won the game.

Question: How does a conman proceed when he takes his opponent to the cleaner's in this game?

A) He tries to name one of the numbers 10, 20, 30, 40, 50, 60, 70, 80, or 90. As soon as he manages this, he proceeds in multiples of ten.

B) He tries to name one of the numbers 2, 13, 24, 35, 46, 57, 68, 79, or 90. As soon as he manages this, he proceeds in multiples of eleven.

C) He keeps naming a number that is 1 more than his opponent's, until his opponent names a number between 80 and 89. Then he responds with 90 and, because his opponent can only say a number between 91 and 100, the conman has won.

2. A force-majeure exercise

Two people playing a game of chance, in which the chances of each player winning a round are exactly fifty-fifty, agree that the first person to win five rounds will receive the entire stake. After the first player has won three rounds and the second player has won one, the game is interrupted by a "force majeure," a superior force.

Question: Based on the rounds played so far, how should the stake be fairly divided up after the game has been interrupted?

A) The stake should be divided up at a ratio of 3:1 in favor of the first player.
B) The stake should be divided up at a ratio of 4:2 (i.e., 2:1), in favor of the first player.
C) The stake should be divided up at a ratio of 13:3 in favor of the first player.

3. A simple game of craps

Two dice are thrown. Beforehand, a player wagers a hundred dollars that the sum of the numbers thrown will be either seven or eleven. If he wins, he receives 300 dollars on top of his stake; if he loses, his stake goes to the gambling house.

Question: What winnings can the gambling house expect after 7,200 such games?

A) The house can expect a profit of about 80,000 dollars (before tax).
B) The house can expect a profit of about 40,000 dollars (before tax).
C) The house cannot expect any profit.

4. The Chevalier de Méré's mistake

The avid gambler Antoine Gombaud, known as the Chevalier de Méré, is said to have come to the following erroneous conclusion: he thought that, when throwing three dice, there was an equal probability of throwing a sum of eleven or a sum of twelve, since the sum of eleven could only be achieved by 1 + 4 + 6 = 1 + 5 + 5 = 2 + 3 + 6 = 2 + 4 + 5 = 3 + 3 + 5 = 3 + 4 + 4 and the sum of twelve could only be achieved by 1 + 5 + 6 = 2 + 4 + 6 = 2 + 5 + 5 = 3 + 3 + 6 = 3 + 4 + 5 = 4 + 4 + 4. That makes six different groups of numbers to get each of eleven and twelve, and, consequently, de Méré's view was that there is an equal probability of throwing a sum of eleven or twelve.

Question: How does this error arise, and which of the two total sums is more probable?

A) The Chevalier de Méré wasn't wrong at all, since the two sums are indeed equally probable.
B) The Chevalier de Méré overlooked the fact that, in the case of the total sum of eleven, three of the possible groups of numbers contain 2 identical summands, whereas in the case of the total sum of twelve, only two of the possible groups of numbers contain 2 identical summands, but another one of the possible groups of numbers is made up of 3 identical summands. The total sum of twelve is more probable than the total sum of eleven.
C) The Chevalier de Méré overlooked the fact that, instead of the group of numbers 1 + 4 + 6 alone, for example, one actually has to take into account the six groups of numbers 1+ 4 + 6, 1 + 6 + 4, 4 + 1 + 6, 4 + 6 + 1, 6 + 1 + 4 and 6 + 4 + 1. If one carries this out with all of the possible groups of numbers, then one comes to the conclusion that the total sum of eleven is more probable than the total sum of twelve.

5. Playing the Paroli system

Let's look at a game of roulette where, for simplicity's sake, the number zero plays no part and one can only bet on evens (such as black or red, where if one wins, one's stake is doubled). As well as the "Martingale" system, whereby one should double one's stake when one loses, there is the "Paroli" system, whereby one doubles one's stake when one wins.

Question: Which of the two systems is more certain: Martingale or Paroli?

A) You should definitely play Martingale: in the long term, you are bound to win.
B) You should definitely play Paroli: at the very most, you can only lose your initial (small) stake.
C) Neither of the two systems is certain.

6. Auctioning off a ducat

A gold ducat (worth more than a hundred dollars) is offered at auction to at least two people. The starting bid is set at a single dollar. However, the following is agreed: the highest bidder will receive the coin from the auctioneer in return for his bid, but the person with the second-highest bid will have to pay the amount of his bid to the auctioneer without getting anything in return. Generally, the auction will proceed as follows: the low bids of a few dollars are quickly surpassed—for every bidder, getting a ducat for a few dollars is good business. The bids quickly reach a value in the region of a hundred dollars. However, even these are outbid, since none of the bidders wants to be the second-highest and have to hand out a lot of money for nothing. As the bids increase, the potential loss becomes more and more painful, and so there is no upper limit to the bidding, apart from that imposed by the bidders' financial reserves. This situation means a huge profit for the auctioneer.

Question: Which strategies could help the bidders to avoid this paradox?

A) If only a few bidders participate in the auction, they could make a binding agreement beforehand.
B) The bidders should not bid beyond a fixed amount.
C) There is no strategy to help the bidders to avoid this paradox.

7. When good intentions go badly

Whether it's the development of a business relationship, the creation of a product, or some other implementation of an idea, one can always talk abstractly of a "path" or "route" that must be taken from a starting point A to a destination Z. Let's imagine the following situation: Two players, called Johnny and Oskar, want to get from A to Z. They have two possibilities: they can either go from A to Z via X or via Y. The route from A to X costs two ducats and the route from X to Z costs five ducats; the route from A to Y costs five ducats and the route from Y to Z costs two ducats. However, there is a catch—each person has to pay double if they travel as a pair. Fortunately, the two players don't have to do this: Johnny goes from A to Z via X and pays seven ducats for this, and Oskar goes from A to Z via Y, also paying seven ducats. So far, so good. But now a fairy, who actually means well, makes it possible for the players to go directly from X to Y without paying a penny. Johnny is delighted to make use of this possibility, while Oskar distrustfully sticks to his normal route. And this is indeed advantageous for Johnny: he only pays two ducats to get from A to X, nothing for the stretch between X and Y and—because he shares the last stretch with Oskar—four ducats to get from Y to Z, making a total of six ducats, one ducat less than before. Oskar, on the other hand, is not so happy, because he now has to pay, on top of the five ducats to get from A to Y, four ducats to get from Y to Z, since Johnny is traveling with him. He therefore has to pay a total of nine ducats, that is, two ducats more than before.

Question: What will Oskar and Johnny end up doing?

A) Oskar and Johnny will return to their original routes.
B) Oskar will stick to his old route, and Johnny will take the new route with the free stretch between X and Y.
C) Both Oskar and Johnny will choose the new route with the free stretch from X to Y—to the detriment of both of them.

8. Escaping the dilemma

Two players are offered the chance to invest in a firm. They can decide individually whether they want to invest or not. If both of them invest, each of them wins two ducats. If only one of the players does so and the other player declines to invest, this means that the player who invests makes a loss of one ducat (the individual investment was too low to put the firm in profit and it goes bankrupt), while the other player, who bet against the firm by declining the investment, pockets four ducats. If both players decline to invest, neither of them makes a profit or a loss. This is a classic example of the prisoners' dilemma: although a joint investment by the players would mean a profit of two ducats for both of them, they will both decline to invest, because it is always better for each player not to invest when acting independently of the other player's decision.

Question: Can the two players find a way out of the prisoners' dilemma if the one who declines to invest receives seven instead of four ducats, assuming that the other player makes the investment and thus loses one ducat?

A) Yes, there is the possibility of the two players making an agreement that is not as tenuous as if they both agreed to invest, and they can escape the prisoners' dilemma with a profit of three ducats each.
B) Only if the two players agree that they will both invest, and they can rely on this agreement being honored, can they

escape the prisoners' dilemma, with a profit of two ducats each.

C) No, even in this case, both players are trapped in the prisoners' dilemma.

9. Left or right?

A hundred-meter beach promenade stretches as straight as an arrow from west to east (on a north-aligned map, from left to right). Before it is the beach, which stretches twenty meters down to the sea along its entire length and is bordered by cliffs at both ends of the promenade. Twenty-five meters to the right of the left-hand end of the promenade and twenty-five meters to the left of the right-hand end of the promenade are the stalls of two ice-cream sellers, which are placed in such a way that the sellers can count exactly half of the sunbathers equally distributed along the beach as their potential customers. One day, the left-hand ice-cream seller thinks to himself, "If I move my stall a few meters to the right, I will gain a few more customers—those in the left-hand section of the beach will still come to me, and I will gain a few from the right-hand section, too." The next day, the right-hand ice-cream seller sees that the other seller has moved and notices at the same time a decrease in his sales.

Question: How will the right-hand ice-cream seller react, what will the left-hand seller then do, and where will the two stalls end up?

A) The right-hand seller will now move much further to the left, thus driving the left-hand seller back to the left, and they will eventually move back to their old spots.

B) The right-hand seller will now also move a few meters to the left, the left-hand seller will move a bit more to the right, and so on, until the two sellers meet in the middle of the promenade.

C) The right-hand seller will now also move a few meters to the left, thus driving the left-hand seller much further to the left.

The right-hand seller will then return to his original spot and the game of the stalls drifting back and forth along the promenade will start again from the beginning.

10. The Monty Hall problem with a difference

Let's assume you are on a game show and have to choose between three doors. Behind two of the doors are cars, the much-coveted prizes of the show, and behind one of the doors is a goat. You choose a door, let's say door number 2, and the show's host, who knows what is behind each door, opens a different door, let's say door number 3, behind which there is a car. He now asks you: "Would you like to choose door number 1?"

Question: Is it advantageous to switch?

A) Yes, it is also advantageous to switch in this case.
B) No, in this case it is better to stick to your first choice.
C) It makes no difference whether you switch or stick to your first choice.

ANSWERS

Answer to 1: B) He must ensure that, in the last round, his opponent says at least 91 and at most 100. He achieves this if he can say 90, and this is possible if his opponent says at least 80 and at most 89 in the penultimate round. The conman achieves this if he can say 79. In this way, he can think back through the rounds of the game and recognize that, as soon as he can say one of the numbers 2, 13, 24, 35, 46, 57, 68, 79, or 90, his victory is assured, because he can then go on to say all the other numbers in this series and thus, after saying 90, force his opponent to say at least 91 and at most 100, after which the conman can reply with at least 101 and has thus won the game.

Answer to 2: C) If A refers to the event whereby the first player wins a round and B refers to the event whereby the second player wins a round, the following series of events come into question if the game is continued, with either A occurring twice with A at the end and B occurring at most three times (good for the first player) or B occurring four times with B at the end and A occurring at most once (good for the second player): AA, ABA, BAA, ABBA, BABA, BBAA, ABBBA, BABBA, BBABA, BBBAA—in these cases, the first player would win the entire stake; the probabilities of these series occurring are 1/4, 1/8, 1/8, 1/16, 1/16, 1/32, 1/32, 1/32, 1/32, making a total of 13/16. Or BBBB, ABBBB, BABBB, BBABB, BBBAB—in these cases, the second player would win the entire stake; the probabilities of these series occurring are 1/16, 1/32, 1/32, 1/32, 1/32, making a total of 3/16. Accordingly, the stake should be divided up between the two players at a ratio of 13:3.

Answer to 3: A) There are 6 x 6 = 36 pairs of numbers, all of which have an equal probability of occurring. Out of all these possible pairs, the only ones that are favorable for the player are the six pairs (1|6), (2|5), (3|4), (4|3), (5|2), and (6|1) that make up a total of seven, and the two pairs (5|6) and (6|5) that make up the total of eleven (i.e., eight pairs in all). The probability of winning is therefore 8/36 = 2/9. This means that the house would have to return the players' stake and pay them 300 dollars on top in approximately 2/9 of the 7,200 games (i.e., to 1,600 players), which means a payout of approximately 1,600 x 300 = 480,000 dollars. For the remaining 7,200 – 1,600 = 5,600 players, the house would keep the stake, which amounts to in total 5,600 x 100 = 560,000 dollars. Accordingly, the house can expect a profit (before tax) of approximately 80,000 dollars.

Answer to 4: C) De Méré would only be right if the three dice were in principle indistinguishable (although it must be pointed out that in quantum theory this case is viewed *mutatis mutandi,* and so in that respect Answer **A** is not completely wrong). If we assume, however, that the three dice are colored blue, green, and red, for example, then a 1 with the blue dice, a 4 with the green dice, and a 6 with the red dice is not the same as a 4 with the blue dice, a 6 with the green dice, and a 1 with the red dice. One must therefore differentiate between (1|4|6) and (4|6|1). This was not taken into account by the Chevalier de Méré in his sums. The following twenty-seven triplets will give the total sum of eleven:

(1|4|6), (1|5|5), (1|6|4),
(2|3|6), (2|4|5), (2|5|4), (2|6|3),
(3|2|6), (3|3|5), (3|4|4), (3|5|3), (3|6|2),
(4|1|6), (4|2|5), (4|3|4), (4|4|3), (4|5|2), (4|6|1),
(5|1|5), (5|2|4), (5|3|3), (5|4|2), (5|5|1),
(6|1|4), (6|2|3), (6|3|2), (6|4|1).

On the other hand, only the following twenty-five triplets will give the total sum of twelve:

(1|5|6), (1|6|5),
(2|4|6), (2|5|5), (2|6|4),
(3|3|6), (3|4|5), (3|5|4), (3|6|3),
(4|2|6), (4|3|5), (4|4|4), (4|5|3), (4|6|2),
(5|1|6), (5|2|5), (5|3|4), (5|4|3), (5|5|2), (5|6|1),
(6|1|5), (6|2|4), (6|3|3), (6|4|2), (6|5|1).

Therefore, the probability of getting a total of eleven (which is $27/216 = 1/8 = 12.5$ percent) is slightly greater than the probability of getting a total of twelve ($25/216 \approx 11.574\%$).

The first part of Answer **B** *is correct*: For three different summands, there are six possible ways to write the combinations, but if exactly two of the summands are identical, there are only three possible ways to write the combinations and if the three summands are identical, there is only one possible way to write the combination. However, in Answer **B**, the statement that the total sum of twelve is more probable than the total sum of eleven is wrong.

Answer to 5: C) Neither of the two systems is certain—if the casino puts in place an upper limit for bets. Without an upper limit, a Martingale player could always gain an amount equivalent to his initial stake as profit when he wins after a series of losses—but this assumes that he has an unlimited amount of money at his disposal. (In that case, Answer **A** would be correct.) By putting in place an upper limit, the casino blocks this possibility and exposes the Martingale player to the *risk of losing a large amount of money*. By contrast, the Paroli player cannot lose any more than the stake he bets at the beginning. On the other hand, he can get much more upset than the Martingale player if he plays one round too many and loses all of the winnings he has accumulated over a number of favorable rounds, plus his initial stake, to the casino. Despite the manageable loss in a *single* Paroli game—only in such a case would Answer **B** be correct—there is a danger that the player, his annoyance overcoming his common sense, wouldn't know when to stop and would keep on playing the Paroli system, which would then create the risk of a major loss, just like in the Martingale system. Fyodor Dostoyevsky, himself addicted to games, describes in his novella *The Gambler* a scene in which a general plays the Paroli system:

> Slowly he took out his money bags, and slowly extracted 300 francs in gold, which he staked on the black, and won. Yet he did not take up his winnings—he left them there on the table. Again the black turned up, and again he did not gather in what he had won; and when, in the third round, the red turned up he lost, at a stroke, 1200 francs. Yet even then he rose with a smile, and thus preserved his reputation; yet I knew that his money bags must be chafing his heart, as well as that, had the stake been twice or thrice as much again, he could not have restrained himself from venting his disappointment.[1]

Answer to 6: A), B), or indeed **C)** If it is possible for the (few) bidders to make a binding agreement among themselves, they could arrange that only one bid of one dollar is made and so the bidder gets the ducat for this price, sells it, and then shares the profit received among all the bidders. If such an agreement is not possible, bidders should on no account bid higher than a fixed amount that they don't mind losing. If the loss of even one dollar is a problem, then one shouldn't get involved in the auction.

Answer to 7: C) Oskar will certainly now also decide to take the free route from X to Y. But Johnny will not return to his former route, because he knows that Oskar will decide to take the route from A via X and Y to Z and that would cost Johnny nine ducats on his old route. So the two of them end up both going along the new route from A to X—cost for each of them: four ducats—from X to Y and Y to Z—which means a further cost of four ducats each. Both of them therefore pay a total of eight ducats each, one more ducat than to begin with. But nothing can induce them not to take the free route from X to Y; they are stuck in a so-called Nash equilibrium dilemma. The fairy's intentions were good in theory, but not in practice. (This paradox, which actually plays an important role in traffic planning and represents a variation of the prisoners' dilemma, was invented in 1968 by the mathematician Dietrich Braess.)

Answer to 8: A) The two players agree that the first one should invest and the second one should decline the investment. The second player gives four ducats of the seven ducats that he wins to the first player. With this strategy, the players win three ducats each—more than if they both made the investment.

Answer to 9: B) The right-hand ice-cream seller will move his stall a few meters to the left in order to combat the threat of losing the potential customers on his left. The left-hand seller will then move further to the right, after which the right-hand seller will drift further to the left, since the left-hand seller considers the customers to his left not to be at risk, with the right-hand seller thinking the same of the customers to his right. Finally, the two stalls will come together in the middle of the beach promenade. This is disadvantageous for both of them, because the potential customers at the very ends of the beach will no longer come to the ice-cream stalls, because they would have to walk almost twice as far as before. (This paradox, a forerunner of the prisoners' dilemma, was invented in

1929 by the American economist Harold Hotelling. The words used for the locations, "left" and "right," can easily be transferred to the political arena: in a country with a strong two-party system, both the left-leaning party and the right-leaning party will try to move toward the middle ground in order to attract potential voters from the other party. The left-wing party will choose somebody from their more right-leaning ranks as their election candidate, while the right-wing party will select one of their more left-leaning members as their candidate. In this way, the parties come closer and closer together and risk becoming interchangeable.)

Answer to 10: B) No, in this case it is better to stick to your first choice. The probability that a car is behind door number 2 is two-thirds. Accordingly, the probability that there is a goat and not a car behind door number 2, and (now that door number 3 has been opened) the second car is behind door number 1, is only one-third. Therefore, with two cars and one goat behind the three doors, it is advisable not to switch.

GLOSSARY

Austrian School of Economics. *See* **Viennese School of Economics**

Bertrand's box paradox. *See* **Monty Hall problem**

Brownian motion: Robert Brown (1773–1858) observed under a microscope how pollen in a drop of water moved erratically. These movements are caused by the thermal movement of the water molecules. Mathematically, this motion was elucidated by Louis Bachelier (1870–1946), Albert Einstein (1879–1955), and Marian von Smoluchowski (1872–1917). Norbert Wiener (1894–1964) developed a sophisticated mathematical theory about it. (See: Andrei N. Borodin and Paavo Salminen, *Handbook of Brownian Motion: Facts and Formulae* [Basel: Birkhäuser, 2002].)

Chance: a necessary condition for **probability theory**. Only if one presupposes the role of chance can the methods of probability theory be applied.

Chicken: A game that involves a test of courage. Two stolen cars are driven at high speed toward each other on a deserted country road. Whichever driver swerves away thus proves that he is a coward and has therefore lost the game. Buzz and Jim, the names given to the players in Nash's dialog with Tucker, are also the names of the two chicken players in the James Dean film *Rebel Without a Cause.* (See: John Maynard Smith, *Evolution and the Theory of Games* [Cambridge, UK: Cambridge University Press, 1982].)

Con game: An apparent game of chance, in which cheating is used to eliminate chance and systematically take money from gullible players. (See: John Nevil Maskelyne, *Sharps and Flats: A Complete Revelation of the Secrets of Cheating at Games of Chance and Skill* [London: Longmans, Green & Co., 1894].)

Deus ex machina. *See* **Force majeure**

Diamond-water paradox: At first sight, it seems odd that a commodity of great use is sometimes traded at a lower price than a commodity of less use.

Dimension: The dimension of a space is essentially the number of possible separate directions in which one can move away from a given point in this space without leaving the space. Luitzen Egbertus Jan Brouwer (1881–1966), Pawel Samuilovich Urysohn (1898–1924), and Karl Menger (1902–1985) defined dimension as "inductive": the outline of a space has a dimension smaller by one than the space itself. Felix Hausdorff (1868–1942) came up with a concept of non-integer dimension. (See: Ryszard Engelking, *Theory of Dimensions: Finite and Infinite* [Lemgo, Germany: Heldermann, 1995].)

Dominant strategy: A strategy that of all the possible strategies promises the greatest benefit for the player. If, of all the possible strategies, only one emerges as the dominant strategy, it is referred to as the game's "strictly dominant strategy."

Double-entry accounting: So called because each business transaction is recorded twofold. It was not the Venetian monk Luca Pacioli (1445–1517) who invented the method of double-entry accounting, as claimed by Pascal in the fictitious dialog, but rather the merchant Benedetto Cotrugli (1416–1469). However, Luca Pacioli was the first to present it in detail.

Drama: Allows people performing on stage to slip into different roles for a certain period and thus to disguise themselves from the audience and perhaps even from themselves. Despite the words written by William Shakespeare (1564–1616)—"All the world's a stage"—the main difference between drama and reality is that, firstly, drama has a clearly defined beginning, secondly, that drama follows a more or less predetermined path, and, thirdly, that it also has a clearly defined end.

Elo rating: A number that provides a rating of a chess player's ability. The concept has in the meantime been adapted for a variety of sports. (See: Arpad E. Elo, *The Rating of Chess Players, Past and Present* [London: Batsford, 1978].)

Force majeure: Meaning "superior force," this applies when circumstances beyond the control of those involved occur, and the event cannot therefore be predicted. In **drama**, this is sometimes put into effect with a "Deus ex machina" plot device—originally meaning the appearance of a god with the help of stage machinery. It uses the unexpected appearance of events, characters, or other external powers to suddenly resolve a conflict.

Game of chance: A game that is predominantly determined by **chance** and can be analyzed with the help of probability theory.

Game of skill: When referring to dexterity games, a certain speed of reaction or coordination in the players is required. Games of skill also include those games that, unlike **games of chance**, require more skill than luck. Chess is one of them, as are bridge and tarock, though poker is considered to be a game of chance.

Game table: Also known as a game matrix, it is a table showing what is paid out to the players in a round of a game. The lines and columns of the matrix represent the possible actions of the players, with the winnings or losses for the players entered in the matrix.

Hex: A strategy board game invented first in 1942 by the Danish poet and mathematician Piet Hein (1905–1996) and given by him the name "Polygon," and then, independently of Hein, in 1947 by John Forbes Nash (1928–2015). (See: Martin Gardner, *Hexaflexagons and Other Mathematical Diversions: The First* Scientific American *Book of Puzzles and Games* [Chicago: University of Chicago Press, 1988].)

***Homo oeconomicus*:** A simplified image of a person who is only interested in a quick profit. (See: Karl Sigmund, *Games of Life: Explorations in Ecology, Evolution and Behavior* [Mineola, NY: Dover, 2017].)

Language game: In the words of Ludwig Wittgenstein (1889–1951), a self-contained communication system. He is referring here to "the whole, consisting of language and the actions into which it is woven."

Law of balance: The **law of large numbers** can easily mislead people into thinking that, after a coin has been tossed many times, with tails coming each time, then the next time the coin is tossed, the probability of heads is greater than half. Naive players make the same false assumption in other games of chance.

Law of large numbers: The frequency with which an event occurs when a game of chance is repeated many times almost always converges with the probability of this event. It must be remembered, however, that there can always be outliers, even with a large number of repetitions, and the "convergence" is not always monotonous.

Lose-lose: The result of a non-zero-sum game in which both players suffer a loss. (See: Roger Fisher and William Ury, *Getting to Yes: Negotiating Agreement without Giving In* [Boston: Houghton Mifflin, 1991].)

Marginalism. *See* **Viennese School of Economics**

Marginal utility: The marginal utility of a commodity refers to the increase in benefit that one experiences by consuming an additional unit of the commodity. The example of the farmer with the five sacks of wheat presented in this book is attributed to Eugen von Böhm-Bawerk (1851–1914).

Martingale: A strategy in **games of chance**, particularly **roulette**, which involves increasing one's stake when one loses. The word comes from the Provençal language and derives from the French town of Martigues, the inhabitants of which were once considered to be rather naive. The Provençal expression "jouga a la martegalo" means "to play very riskily." (See: Lester E. Dubins and Leonard J. Savage, *How to Gamble if You Must: Inequalities for Stochastic Processes* [New York: McGraw-Hill, 1965].)

Maximin rule: A strategy following this rule guarantees the player the highest possible winnings that can be achieved regardless of what the other player does. If both players choose their strategy according to the maximin rule, their game remains stuck in a **Nash equilibrium**.

Méziriac's think-of-a-number game: A combinational game that turns out to be a **con game**.

Monty Hall problem: Also known as Bertrand's box paradox after Joseph Bertrand (1822–1900). In Bertrand's version, there are three boxes, in one of which is a gold coin. The player chooses a box, puts it to one side, and opens one of the two remaining boxes. He sees that this box is empty. What is the probability that the box laid aside contains the gold coin? Monty Hall's three-door game on the TV show *Let's Make a Deal* is based on the same problem, but behind the doors are goats and a car.

Moore's paradox: The English philosopher George Edward Moore claimed that it would be absurd to say a sentence such as the following: "It is raining but I believe that it is not raining."

Nash equilibrium: Refers to a combination of strategies of the two players in a game, whereby each player chooses his strategy in such a way that it doesn't make sense for either of them to depart from his chosen strategy. (See: Harold W. Kuhn and Sylvia Nasar, eds., *The Essential John Nash* [Princeton: Princeton University Press, 2007].)

Poker: The name of a card game played with an Anglo-American deck of fifty-two cards, with each hand made up of five cards. Without knowing what cards the opponent holds, the players make a bet, of differing amounts, on the chances of their own hand winning. The money thus staked by the players goes to the player with the strongest hand, or to the only player who is left if all the other players are not prepared to match the bet he has made. This makes it possible to win by bluffing, even if one holds weak cards. (See: David Sklansky, *The Theory of Poker* [Henderson, NV: Two Plus Two, 2007].)

Prisoners' dilemma: A game in which the two players can either work together or betray each other. Both players must decide their strategy without knowing what the other player does. It is therefore possible that one player does the opposite of what the other decides, in which case only the player who "betrays" the other one benefits and thus gains a greater profit. The names "Al" and "Capone" used by Albert Tucker in the lecture described were originally used in this context by Alexander Mehlmann (born 1949). (See: Robert Axelrod, *The Evolution of Cooperation* [New York: Basic Books, 1984].)

Probability theory: A branch of mathematics in which elements of a sample space are termed "elementary events" and groups of elementary events are called "events." Each elementary event is assigned a positive figure, the "probability" of its occurrence. The probability of an event is the sum of the probabilities of the elementary events comprised within the event. The probabilities of all the elementary events are so determined in such a way that their total sum amounts to 1, or 100 percent. The way in which this abstract mathematical theory thought up by Andrey Nikolaevich Kolmogorov (1903–1987) is applied to concrete examples is a "practical" issue, rather than a mathematical one. The typical application examples are: tossing a coin—the elementary events are "heads" and "tails," with the probability of each of these being 1/2; rolling a die—the elementary events are the numbers from one to six, with the probability of each of these being 1/6; roulette—the elementary events are the numbers from zero to thirty-six, with the probability of each of these being 1/37. (See: William Feller, *An Introduction to Probability Theory and Its Applications* [New York: John Wiley & Sons, 1968].)

Quid pro quo: Also known as "tit for tat," this refers to a game strategy in a game conducted over many rounds with no foreseeable end, with a possible **win-win** situation, in which one bets on win-win in the first round and then, in the following rounds, bets exactly as the other player has done in the previous round. In the variation known as "Tit for Two Tats," one doesn't bet on win-win if the other player has not done so in the previous two rounds. (See: Karl Sigmund, *Games of Life: Explorations in Ecology, Evolution and Behavior* [Mineola, NY: Dover, 2017].)

Risk: The product of the probability that an unwanted event occurs and the harm caused by this unwanted event. (See: Nicholas Rescher, *Risk: A Philosophical Introduction to the Theory of Risk Evaluation and Measurement* [Washington, DC: University Press of America, 1983].)

Roulette: In roulette, players bet on numbers between zero and thirty-six or on groups of numbers. The ball spins on the wheel and lands by chance in a slot, thus determining which number comes.

St. Petersburg paradox: Refers to a game of chance in which one can, on average, count on infinitely large winnings, although one only bets a finitely large stake. (See: Karl Menger, "The Role of Uncertainty in Economics," in *Selected Papers in Logic and Foundations, Didactics, Economics* [Dordrecht, Holland: D. Reidel, 1979].)

Strategy game: A game in which a long-term plan on how to proceed in the game is decisive. It is related to the **tactical game**, where the player has to make situational decisions in order to gain a significant advantage. Strategy games can contain elements of luck or chance, though this is not necessary.

Strictly dominant strategy. *See* **Dominant strategy**

Tactical game: A game in which the players have to make situational decisions in order to gain a significant advantage in the game. It is related to the **strategy game**, where a long-term plan on how to proceed is decisive. Tactical games generally contain elements of luck or chance.

Tit for tat. *See* **Quid pro quo**

Tit for Two Tats. *See* **Quid pro quo**

Trend: A measure of how the valuation of a commodity changes over time. Trend research focuses on observing and predicting trends.

Tulip mania: The first well-documented speculative bubble in economic history. It occurred during the Dutch Golden Age in the first half of the seventeenth century, when tulips bulbs became a speculative commodity.

Ultimatum: A game to evaluate to what extent the player is prepared to place his own profit at a level proportionate to that of his fellow player. (See: Steven D. Levitt and Stephen J. Dubner, *Superfreakonomics: Global Cooling, Patriotic Prostitutes and Why Suicide Bombers Should Buy Life Insurance* [New York: HarperCollins, 2009].)

Vienna Circle: A group of philosophers and scientific theorists who gathered in a lecture hall in the building housing the University of Vienna's departments of mathematics, physics, and chemistry every Thursday between 1922 and 1936, led by Hans Hahn (1879–1934) and Moritz Schlick (1882–1936). Those who were part of or related to the Vienna Circle were Alfred Jules Ayer (1910–1989), Gustav Bergmann (1906–1987), Rudolf Carnap (1891–1970), Herbert Feigl (1902–1988), Philipp Frank (1884–1966), Kurt Gödel (1906–1978), Olga Hahn-Neurath (1882–1937), Carl Gustav Hempel (1905–1997), Victor Kraft (1880–1975), Karl Menger (1902–1985), Heinrich Neider (1907–1990), Otto Neurath (1882–1945), Willard Van Orman Quine (1908–2000), Theodor Radakovic (1895–1938), Hans Reichenbach (1881–1953), Alfred Tarski (1902–1983), and Friedrich Waismann (1896–1959).

Viennese School of Economics: Also known as the Austrian School of Economics or marginalism, its leading figures are Carl Menger (1840–1921), Eugen von Böhm-Bawerk (1851 -1914), Ludwig von Mises (1881–1973), Friedrich von Hayek (1899–1992), Murray Rothbard (1926–1995), and Israel Kirzner (born 1930).

Wheel-watcher: A person who claims to know, by observing the spin of the ball and the speed at which the roulette wheel turns, which number the ball will fall on.

Win-win: The result of a non-zero-sum game in which both players profit. (See: Roger Fisher and William Ury, *Getting to Yes: Negotiating Agreement without Giving In* [Boston: Houghton Mifflin, 1991].)

Zero-sum game: A game in which the sum of the winnings and losses of all players amounts to zero. If there are only two players in a zero-sum game, then one of the players loses exactly what the other wins.

NOTES

CHAPTER 1: PLAYING WITH WATER AND DIAMONDS

1. Felix Somary, *Erinnerungen Aus Meinem Leben* [The Raven of Zürich: The Memoirs of Felix Somary] (Zurich: Manesse Verlag, 1955).

CHAPTER 3: PLAYING WITH NUMBERS

1. Blaise Pascal, *Pensées* [Thoughts], 1669, section 2:131.
2. Johannes Kepler to Baron von Herberstein, letter, May 15, 1596, in Carola Baumgardt, *Johannes Kepler: Life and Letters* (New York: Philosophical Library, 1951), p. 22.

CHAPTER 6: PLAYING WITH A SYSTEM

1. Rolf Dobelli, *The Art of Thinking Clearly* (New York: HarperCollins, 2013), p. 86.

CHAPTER 7: PLAYING WITH SCHOLARS

1. Ludwig Wittgenstein, *Tractatus Logico-Philosophicus*, trans. C. K. Ogden (London: Kegan Paul, Trench, Trubner & Co., 1922), 6.45.
2. Ibid., preface.

CHAPTER 9: PLAYING WITH LIFE AND DEATH

1. Herbert A. Simon, review of *Theory of Games and Economic Behavior*, by John von Neumann and Oskar Morgenstern, *American Journal of Sociology* 50, no. 6 (May 1945): 558–60.

2. Arthur H. Copeland, review of *Theory of Games and Economic Behavior*, by John von Neumann and Oskar Morgenstern, *Bulletin of the American Mathematical Society* 51, no. 7 (July 1945): 498–504.

3. Leonid Hurwicz, "The Theory of Economic Behavior," *American Economic Review* 35, no. 5 (December 1945): 909.

4. Jacob Marschak, "Neumann's and Morgenstern's New Approach to Static Economics," *Journal of Political Economy* 54, no. 2 (April 1946): 97–115.

5. Steve J. Heims, *John von Neumann and Norbert Wiener: From Mathematics to the Technologies of Life and Death* (Cambridge, MA: MIT Press, 1980), p. 370.

CHAPTER 10: PLAYING WITH CHICKENS AND LIONS

1. Sylvia Nasar, *A Beautiful Mind: The Life of Mathematical Genius and Nobel Laureate John Nash* (New York: Simon and Schuster, 1998), p. 71.

CHAPTER 12: PLAYING WITH PROFIT

1. Anatol Rapoport, *Certainties and Doubts: A Philosophy of Life* (Montreal: Black Rose Books, 2000), p. 2.

2. Ibid., pp. 2–3.

3. Ibid., p. 3.

CHAPTER 14: PLAYING WITH INFORMATION

1. "Game Show Problem," marilynvossavant.com, http://marilynvossavant.com/game-show-problem.

2. Ibid.

3. Ian Stewart, review of *A Budget of Trisections*, by Underwood Dudley, *Mathematical Intelligencer* 14, no. 1 (Winter 1992): 73–74.

CHAPTER 15: PLAYING WITH LANGUAGE

1. R. B. Braithwaite, "George Edward Moore, 1873–1958," in *G. E. Moore: Essays in Retrospect*, ed. Alice Ambrose and Morris Lazerowitz (London: Routledge, 1970), p. 30.

2. Ludwig Wittgenstein, *Tractatus Logico-Philosophicus*, trans. C. K. Ogden (London: Kegan Paul, Trench, Trubner & Co., 1922), 6.54.

3. Ludwig Wittgenstein, *Philosophical Investigations*, trans. G. E. M. Anscombe (Oxford, UK: Basil Blackwood, 1958), p. 47, sec. 109.

CHAPTER 16: PLAYING WITH EMOTIONS

1. Alexander Mehlmann, *The Game's Afoot! Game Theory in Myth and Paradox*, trans. David Kramer (Providence, RI: American Mathematical Society, 2000), pp. 133–34.

2. Albert Costa et al., "Your Morals Depend on Language," *PLoS One* 9, no. 4 (2014), http://journals.plos.org/plosone/article?id=10.1371/journal.pone.0094842.

3. Mehlmann, *Game's Afoot!*, p. 87.

NUMBER GAMES

1. Fyodor Dostoyevsky, *The Gambler*, trans. C. J. Hogarth (London: J. M. Dent, 1915), ch. 2.

INDEX

Agamemnon, 185–86
Alexandrov, Pavel Sergeyevich, 36
Andermann, Hermine, 16
Annus Mirabilis of Albert Einstein, 34–35
Anselm of Canterbury, 89
architect of the world, 44–45
Aristotle, 43
"Arithmetic of Morality, The," 154
Ars Conjectandi (*The Art of Conjecturing*) (Jakob Bernoulli), 56, 75
Artin, Emil, 119–20
Art of Thinking Clearly, The (Dobelli), 74–75
Ask Marilyn (column), 165, 167–72, 173
As You Like It (Shakespeare), 185
atomic bombs, 115, 116, 120
atoms and molecules, 85
Axamit, Hilda "Mitzi," 84
Axelrod, Robert, 144

Bachelier, Louis, 33–34
Bachet de Méziriac, Claude Gaspard, 37–40, 203
bank robbery. *See* prisoners' dilemma
beach ice-cream vendor problem, 209–10, 214–15
Beaumarchais, Pierre, 189
Belcredi, Richard von, 14
Bernoulli, Daniel, 75
Bernoulli, Jakob, 56, 75–79
Bernoulli, Johann, 56
Bernoulli, Nikolaus, 75–79
Bethe, Hans, 116
"Big Game, The," 172–73
bluffing in poker, 107–108
Bohr, Niels, 44
Boltzmann, Ludwig, 44, 84–85
bond prices. *See* stock market fluctuations
Bonno, Giuseppe, 189
Book on Games of Chance, The (Cardano), 49–50
Borel, Émile, 111
Braithwaite, Richard Bevan, 180–83
bridge, 41, 171–72
Brouwer, Luitzen Egbertus Jan
    dimension theory and, 35–36
    falling out with Menger, 36
    lecture series participation, 91, 176–77
    Menger working for, 23
Brown, Robert, 33
Brownian motion, 33–34, 35

Café Philosophique, 153–54
Cardano, Gerolamo, 49–50
card games, 41
"carrosses à cinq sols," 195–96
Čech, Eduard, 98
chance, 43, 44, 50–55
chess, 41–42

Chevalier de Méré. *See* Gombaud, Antoine
chicken game, 122–29, 135–36
coin flipping, 198–202
Cold War, 151
con games
  curves and, 42–43
  force majeure problem and, 54–55
  think-of-a-number game as, 40
  as zero-dimensional, 112
Copeland, Herbert, 114
Copernicus, 43
Così fan tutte (Mozart), 192–93
cosinus hyperbolicus, 131
Costa, Albert, 187
craps, 204, 211
Cuban Crisis, 129
curves
  dimensions of, 31–32
  prediction and, 40–41, 42–43
Cypria (attributed to Stasinus), 185–86

Dan, Klara, 113
Dance of the Photons: From Einstein to Quantum Teleportation (Zeilinger), 44
Degenfeld-Schönburg, Ferdinand, 97
de Méziriac, Claude Gaspard Bachet, 37–40, 203
Dettonville, Amos, 69
diamond-water paradox (paradox of value), 14–15
dice, 55–56
  game of, 59–64
differential calculus, 44
dimensions of curves, 31–32
dimension theory, 34–36
Dimension Theory (Karl Menger), 23, 32, 36
Dobelli, Rolf, 74–75
Doyle, Arthur Conan, 95. *See also* Sherlock Holmes train dilemma
Dresher, Melvin, 131
ducat auction problem, 206–207, 214
Duffin, Richard, 119

Egorov, Dmitri Fyodorovich, 35
Einstein, Albert
  Annus Mirabilis of, 34–35
  on architect of the world, 44, 45
  Brownian motion paper, 34
  Nash and, 121
  Pauli and, 18
Eisenhower, Dwight D., 115–16
Elo, Arpad Emrick, 41–42
Elo ratings, 41–42
Encyclopedia of Mathematical Sciences, Pauli in, 18
espionage, 172–73
Evolutionary Games and Population Dynamics (Sigmund and Hofbauer), 156
existence of God, 89, 117, 198–202

fair dice, 55–56
Faust (Goethe), 45–46
Fermat, Pierre de, 52–54
Final Problem, The (Doyle), 95. *See also* Sherlock Holmes train dilemma
folding in poker, 107–108
force majeure problem, 50–55, 204, 210–11

France, 39
Frank, Philipp, 86
Franklin, Benjamin, 57–59
Franz Joseph (emperor), 16
Frederick (king), 14
Fuld Hall meetings, 122
Furtwängler, Philipp, 21

game of dice. *See* dice: game of
game of information, 172–73
games of chance
    as one-dimensional, 111–12
    parlor games compared, 112
    zero-dimensional, 112
    as zigzag lines, 43, 44
    See also specific games
Games of Life (Sigmund), 149
game theory, economic behavior explained by, 130
God, existence of, 89, 117, 198–202
Gödel, Kurt
    Furtwängler and, 21
    Hahn impressed by, 13
    lecture series participation, 91
    move to US, 113
    in Vienna Circle, 87, 89
Goethe, 45–46
Golden Gate Bridge, 131
Gombaud, Antoine (Chevalier de Méré)
    with Pascal, 47–50, 195–202
    roulette system and, 70–74
    three-dice mistake, 205, 211–12
Gouffier, Artus, 195
"Great Game, The" 172–73
Grebber, Jan, 65–67
Gröller, Michael, 59
Gunn, Glenn Dillard, 150

Habsburg Empire, 13–14
Hahn, Hans
    Brouwer lecture and, 91
    death, 97
    expanding thought of, 85–86
    in lecture series, 90
    as math teacher, 21–22
    on Menger's book, 83, 92–93
    nature of curve and, 21–23, 25–27
    support of Menger by, 13, 23
    Vienna Circle and, 86–87
Hall, Monty, 166–67
Halparin, Monte, 166
happiness, 200–201
Harsanyi, John (János), 111
Harvard University, 156–57
Hausdorff, Felix, 22–23, 31
Heisenberg, Werner, 90
Hlawka, Edmund, 156
Hofbauer, Josef, 156
Hofreiter, Nikolaus, 155
horse-drawn omnibuses, 195–96
Hosch, Reinhard, 153–54, 159
Hurwicz, Leonid, 114, 115

information game, 172–73
Institute for Advanced Study, 98, 113
insurance, 61–65
integral dimensions, 32
investment game
    explained, 136–38
    prisoners' dilemma and, 136–38, 140, 208–209, 214
    See also iterated investment game
iterated investment game
    explained, 142–49

human nature and, 149
police version of, 158–60
praise versus punishment and, 159–60
Sigmund lecture on, 154, 157–60
taxes and, 161–62

Johnny and Oskar paths problem, 207–208, 214
Joseph, grain prices and, 59

Kepler, Johannes, 43–45
Keynes, John Maynard, 175
Korean War, 115
Kövesi, Mariette (later von Neumann), 113
Kuhn, Richard, 18

La Folle Journée: ou le Mariage de Figaro (Beaumarchais), 189
language games
moral dilemmas affected by, 187–88
Pascal and, 198–202
Popper and, 182–83
Wittgenstein on, 178–80
Law, John, 15
law of balance, 74–75
law of large numbers, 56, 60
Lefschetz, Solomon, 120
left or right problem, 209–10, 214–15
Lem, Stanisław, 45
Let's Make a Deal, 166–67
Lettres Provinciales (Montalte), 69
Levy, Paul, 34
loaded dice, 55–56
lotteries, reason for popularity of, 64
Luzin, Nikolai Nikolaevich, 35

MacArthur, Douglas, 115
Mach, Ernst, 84–85
Manchmal gewinnt der Bessere: Die Physik des Fussballspiels (Sometimes the better team wins: The physics of soccer) (Tolan), 42
"Manus manum lavat" (maxim), 141, 145. *See also* iterated investment game
marginal use, 16, 82
Marguerite, eye healing, 196–97
Maria Theresia (empress), 14
Mark, Hermann, 90
Marriage of Figaro, The (Mozart), 189–92
marriage sui juris, 16–17
Marschak, Jacob, 114
Martingale system for roulette, 70–74, 206, 212–13
"Mathematical Colloquium," 90
maximin rule, 107, 108, 115
Mayer, Hans, 97
Mehlmann, Alexander, 185–86, 188
Mémorial of Blaise Pascal, 197–98
Menger, Carl
background of, 13–14
birth of son and, 16–17
career, 14, 16–17
death, 84
diamond-water paradox and, 14–15
marginal use and diamond-water paradox, 16
as melancholic, 17
Menger, Karl
birth of, 16–17
Brouwer lecture and, 91
childhood of, 17–18
correspondence of, 84

desire to return to Vienna, 154–55
Dimension Theory, 23, 32, 36
drama interest of, 19
falling out with Brouwer, 36
health of, 83–84
influences on, 84–85
leaving Vienna, 97
lecture series organizer, 90
mathematics, switch to, 20–23, 26
Morality, Decision and Social Organization, 83, 92–93, 95–96
physics interest of, 18–19
St. Petersburg paradox and, 82
support by Hahn of, 13, 23
in Vienna Circle, 87, 89–90
von Neumann compared, 110–11
WWII and, 91
metaphysics, 88
Milnor, John, 129
Minor, Eleonore, 22
molecules and atoms, 85
Montalte, Louis de, 69
Montmort, Pierre Rémond de, 75
Moore, George Edward, 177, 178
Moore's paradox, 178
moral dilemma with train, 187–88
Morality, Decision and Social Organization (Karl Menger), 83, 92–93, 95–96
Morgenstern, Oskar, 93, 97–98, 113–14. *See also* Sherlock Holmes train dilemma
Mozart, Wolfgang Amadeus, 189–92
music, 188–92, 193

Nash, Alicia, 129
Nash, John
chicken game and, 122–29
Einstein and, 121
Feld Hall meetings and, 122
meeting Artin, 119–20
mental health issues, 129–30
von Neumann's theory and, 128–29
natural science, history of, 43–44
Neumann, Max, 110
Neurath, Otto, 86, 87–88
Neu-Sandez in Galicia, 13–14
Nobel Prize winners, 18
Nowak, Martin, 156

Odysseus, 185–86
"Old Problems—New Solutions in the Exact Sciences" lecture series, 90–91
ontological argument, existence of God, 89, 117, 198–202
operas, 189
Oppenheimer, Robert, 120
Orsini-Rosenberg, Franz, 189–92
Oskar and Johnny paths problem, 207–208, 214

Pacioli, Luca, 49–50, 218
Palamedes, 185–86
paradox of value (diamond-water paradox), 14–15
parlor games, 112–15
Paroli system for roulette, 206, 212–13
Pascal, Blaise
academy member, 52
Fermat and, 52–54
force majeure problem, 50–56

on freedom, 39
with Gombaud, 47–50, 195–202
health of, 196
horse-drawn omnibuses and, 195–96
Mémorial of, 197–98
at Port Royal des Champs, 47
pseudonyms of, 69
roulette system of, 70–74, 206
St. Petersburg paradox and, 80–82
Pascal's Wager (existence of God), 89, 117, 198–202
paths problem, 207–208, 214
Pauli, Wolfgang, 18
Pauli exclusion principle, 18
Perfect Vacuum, A (Lem), 45
Périer, Gilberte, 195
Phantom of Princeton. *See* Nash, John
Planck, Max, 87
Poincaré, Henri, 33–34
points of equilibrium, 126–27, 128
poker
bluffing, 107–108
novelty of von Neumann and, 111
with two cards, 95, 100–108
as two-dimensional, 112
Popper, Karl, 180–83
Port Royal des Champs, 47
praise versus punishment, 159–60
prediction, curves and, 40–41, 42–43
Presidential Medal of Freedom for von Neumann, 115–16
Principia Mathematica (Russell and Whitehead), 85–86
Principles of Economics (Menger and Menger), 84
prisoners' dilemma
chicken game compared, 135–36
explained, 132–36
investment game compared, 136–38, 140, 208–209, 214
Tosca and, 188
probability
of dice, 55–56
in *force majeure* problem, 52–56
seeing the future versus, 59
three-door game and, 167–72
time affecting, 65–67
See also dice: game of; roulette
Problèmes plaisants et délectables, qui se font par les nombres (Pleasant and delectable problems, which are made by numbers) (Bachet de Méziriac), 39, 203
Program for Evolutionary Dynamics (Harvard), 156–57
Puccini, Giacomo, 188
punishment versus praise, 159–60

quantum phenomena, 44
quid pro quo, 141, 145. *See also* iterated investment game

Radon, Johann, 155–56
Ramsey, Frank Plumpton, 177
RAND Corporation, 115, 129
ranking lists, 41–42
Rapoport, Anatol, 149–51
iterated investment game program and, 145–49
return to Vienna of, 152
Rashevsky, Nicolas, 150–51
Reidemeister, Kurt, 87–89
Riesenhuber, Karl, 153–54

right or left problem, 209–10, 214–15
Rothschild, Louis Nathaniel de, 59
roulette
  explained, 69–70
  law of balance and, 74–75
  Paroli system for, 206, 212–13
  Pascal's system for, 70–74, 206, 212–13
  reason for popularity of, 64
  screens showing previous spins, 171
  St. Petersburg paradox and, 75–82
Rudolf (crown prince), 16
Russell, Bertrand
  Hahn and, 86, 87
  Principia Mathematica, 86
  Wittgenstein and, 177, 180–81, 182–83

saddle points, 106–107, 111, 128
Sautet, Marc, 154
Schlick, Moritz, 86–87, 90–91, 97
Schmetterer, Leopold, 156
Schmoller, Gustav, 17
Schnitzler, Arthur, 19
Schnitzler, Heinrich, 19
Schreier, Otto, 22–23, 31–32
Shakespeare, William, 185
share/bond prices. *See* stock market fluctuations
Sherlock Holmes train dilemma, 95–96, 98–100
Siegel, Carl Ludwig, 156
Sigmund, Karl
  evolution and, 149
  Evolutionary Games and Population Dynamics, 156
  morality lecture, 154, 157–60
  Schmetterer as teacher of, 156
  Ultimatum game, 162–64
Simon, Herbert A., 114
Smith, Adam, 14–15
Smith, John Maynard, 149
Smoluchowski, Marian, 35
Somary, Felix, 17
"Sometimes Tit for Tat" (program), 148
Spann, Othmar, 97
spin doctors, 172
spying, 172–73
Stasinus, 185
statistical physics, 44
Sterne, Laurence, 29
Stewart, Ian, 168
stock market fluctuations
  as curve, 27–31
  impacts on, 32–33
  Menger versus Bachelier approach to, 34
St. Petersburg paradox, 75–82
Strittmatter, Anselm, 117
supply and demand, grain prices and, 59
Szegő, Gábor, 111

tarock, 172
Taussky, Olga, 22, 90
taxes, impact of, 161–62
Teller, Edward, 116
Théorie de la Roulette, La (Dettonville), 69
Theory of Games and Economic Behavior (Morgenstern and von Neumann), 113–15
think-of-a-number game, 37–41, 203, 210

Thirring, Hans, 90
three-dice mistake, 205, 211–12
three-door game
  on *Let's Make a Deal*, 166–67
  Marilyn vos Savant answering question about, 167–72
  problem about, 210, 215
"Time is money" (phrase), 57–59
tit for tat, 141, 145. *See also* iterated investment game
"Tit for Two Tats" (program), 148
Tolan, Metin, 42
Tractatus logico-philosophicus (Wittgenstein), 87–89, 177, 179
train dilemma. *See* Sherlock Holmes train dilemma
train moral dilemma, 187–88
treble clef, 29–30, 31
tree diagram. See *force majeure* problem
Tristram Shandy (Sterne), 29
Truman, Harry, 115
truth, 139–40
Tucker, Albert
  chicken game and, 122–29
  economic behavior and game theory, 130
  investment game and, 136–38
  prisoners' dilemma and, 132–36
  in San Francisco, 131
tulip mania, 65–67
two-ity, 176

Ulam, Stanley, 113
Ultimatum (game), 162–64
University of Vienna, 17, 86–87, 97, 154–55
Urysohn, Pavel Samuilovich, 35, 36

value in exchange, 14–15
value in use, 14–15
van Swieten, Gottfried, 189, 191–92
Vetsera, Mary, 16
Vienna, impact of WWI on, 19–20
Vienna Circle, 86–90, 92, 97
von Mises, Richard, 85–86
von Neumann, John (János)
  book with Morgenstern, 113–14
  death, 116–17
  at Institute for Advanced Study, 98
  Menger compared, 110–11
  moving to US, 112–13
  Nash's theory and, 128–29
  parties, 113
  RAND Corporation and, 115
  staying in US, 109
  See also Sherlock Holmes train dilemma
von Neumann, Mariette (née Kövesi), 113
von Strack, Johann, 189
vos Savant, Marilyn, 165, 167–72, 173

Waismann, Friedrich, 88–89
wheat and water paradox, 15–16
wheel-watchers, 171
Whitaker, Craig F., question from, 165, 167–72, 173
Whitehead, Alfred North, 85–86
Wigner, Jenő (Eugene), 111
Wirtinger, Wilhelm, 20–21
Wissenschaftliche Weltauffassung (Scientific World-Conception), 91–92
Wittgenstein, Ludwig
  arrival in England, 175
  Brouwer lecture and, 90–91, 176–77

doctoral thesis, 177–78
Keynes on, 175
on language, 178–80
on Moore's paradox, 178
Popper debate and, 180–83
stage of the world and, 187
Tractatus, 87–89, 177, 179
world stage, 187, 188
World War I, 19–20
World War II, 91, 92, 97–98, 112–13

Zeilinger, Anton, 44
zero-sum games, 108, 130, 147, 162
zigzags, games of chance and, 43, 44. *See also* curves
Zweig, Stefan, 17